Axel Goës

A Synopsis of the Arctic and Scandinavian Recent Marine Foraminifera

Hitherto Discovered

Axel Goës

A Synopsis of the Arctic and Scandinavian Recent Marine Foraminifera
Hitherto Discovered

ISBN/EAN: 9783337139995

Printed in Europe, USA, Canada, Australia, Japan

Cover: Foto ©berggeist007 / pixelio.de

More available books at **www.hansebooks.com**

A

SYNOPSIS

OF THE

ARCTIC AND SCANDINAVIAN

RECENT MARINE

FORAMINIFERA

HITHERTO DISCOVERED

BY

AXEL GOËS.

WITH TWENTY-FIVE PLATES.

COMMUNICATED TO THE R. SWEDISH ACADEMY OF SCIENCES OCTOBER 12, 1892.

STOCKHOLM 1894.
KUNGL. BOKTRYCKERIET. P. A. NORSTEDT & SÖNER.

This Synopsis is founded on materials extant in the Swedish State Museum and for the most part brought together by the various arctic expeditions which during the two decennaries 1858—1878 were fitted out from Sweden, as also by the dredgings of numerous naturalists which for many years have investigated its shores.

A brief review of the cruises and dredging-excursions thus carried out, will give the reader some notice about the stations searched.

In the year 1858 OTTO TORELL, accompanied by ADOLF NORDENSKIÖLD and AUG. QVENNERSTEDT, undertook a private cruise of two months on the western parts of Spitzbergen, and brought home extensive marine and geological collections, particularly from Icesound, Bellsound and Hornsound, where they reached the greatest depth at 900 meters.

The arctic expedition of 1861, with O. TORELL as leader, resulted also in rich collections, made by ANDREW MALMGREN, ADAM FREDERIC SMITT, and myself. Extensive dredgings were undertaken in the bays and surrounding seas of the western, northern, and eastern parts of Spitzbergen, and the late Dr. KARL CHYDENIUS from Helsingfors made deep sea soundings in 75° n. lat. and 12° east long., where the bottom was caught in 1,869 meters, and in 76° 17′ n. lat., 13° 53′ east. long., where the Bulldog machine struck bottom at a depth of 2,490 meters.

In 1864 a reconnoitering expedition under the command of NORDENSKIÖLD, accompanied by MALMGREN as zoologist, visited the southern bays of Spitzbergen and brought home valuable collections.

In 1868 the steamer "Sophia", commander now Admiral Baron F. W. VON OTTER, was sent to the seas of Spitzbergen on an expedition the scientific chief of which was NORDENSKIÖLD, while SMITT and MALMGREN were attached as zoologists. Numerous dredgings were executed and deep sea soundings made, so in 81° 32′ n. lat. northwest from Spitzbergen, where the sounding machine brought up bottom clay from 2,314 meters, and in 78° 26′ n. lat., 2° 17′ west. long., from 4,700 meters, besides at other stations west and northwest of Spitzbergen from depths of 1,000—2,000 meters.

During a reconnoitering tour to Greenland in 1870 NORDENSKIÖLD was accompanied, amongst other men of science, by Dr. P. ÖBERG who at several places in Baffins Bay brought up quantities of typical arctic Foraminifera of high development, from a depth of 100 to 650 meters. The principal stations were Jacobshavn, Claushavn, Tossukatek, Disco.

Dr. Josua Lindahl, who previously had assisted at the scientific cruise of the "Porcupine", had the fortune in 1871 to join H. M. gunboats "Ingegerd" and "Gladan", on their visit to Greenland. He made deep-dredgings with the best result in the north-western Atlantic, and in Baffins Bay up to 72° n. lat. In 53° 34' n. lat. and 52° west. long. he got some hauls from 1,780 meters.

In the year 1875, when Nordenskiöld set out for the Kara Sea, Nova Zembla and the Yenisej river, he was joined by Hjalmar Théel and Anton Stuxberg. In 1876 and 1878 about the same parts of the Arctic Sea were again visited, the last year by the renowned Expedition of the "Vega". The greatest depth in these seas seems to be about 270 meters.

As for the Scandinavian seas they have been perseveringly searched by several of our ablest zoologists. Sven Lovén who was the first who undertook systematically to explore our western shores already in the year 1831, has largely contributed to our knowledge of the fauna of Evertebrates from these parts of the North-Atlantic.

In 1865 Dr. Axel Ljungman made a successful dredging tour to the Koster Islands in the Skagerack, and in 1889 I visited these islands myself, both voyages resulting in a good harvest of Foraminifera. The greatest depth in this part of the Skagerack amounts to about 220 meters.

On a surveying cruise of H. M. gunboat "Gunhild" in 1877 Hj. Théel and Carl Bovallius made dredgings with the best result in the Skagerack to the depth of 900 meters. In the two following years the same work was continued by Théel, Filip Trybom and C. Forsrhand, along the coasts of the southern and western provinces of Sweden.

During the past decennium important contributions to the collections have been made by Carl Aurivillius and Axel Wirén, which on their stays at the biological station of the Swedish Academy of Sciences on our western coast, have made interesting finds particularly in the bay of Gullmaren, where a well developed Rhizopodal fauna is met with in 150 meters.

In the deep bays of Norway, as in the fjords of Qvænangen and Hardanger, Carl Aurivillius with his usual sedulity and success performed extensive dredgings in 1884 and 1888. But for our acquaintance with the Rhizopod fauna of these places we are mainly indebted to the eminent English naturalist Reverend Dr. M. Norman, who in 1883 with the greatest liberality presented to our State Museum an exquisite collection from various Norwegian localities, which he had visited several times. From the surroundings of Bergen in Norway Wilh. Lillieborg and Gustaf Lindström carried home some materials which have yielded many interesting forms.

It is due also here to recall the work done by Rupert Jones and W. Kitchen Parker which, as early as in 1857 and 1865, had largely contributed to our information regarding the North-Atlantic Rhizopodal fauna; as also that of Mr. Henry Brady in several papers on the finds of the last English and Austrian arctic expeditions, reproduced in his great work on the Foraminifera in the Challengers Report 1884.

For the sake of a completing comparison some extraneous forms have also been introduced in this synopsis, as from Smitts and Ljungmans fine collections from the Atlantic off the Azores, and from my own finds in the Caribbean Sea.

Although, in this way, many and various localities have been searched, it is still not to be expected, that our collections at present should afford materials enough for a somewhat complete synopsis and study of the fauna in question. Many a time the whole material from a locality consisted of a small quantity of sand or mud accidentally adhering to the more conspicuous contents of the dredge, and only occasionally I have been enabled to examine somewhat larger quantities of bottom samples. Particularly amongst the Astrorhizidæ and Trochamminidæ, as also in the family of Lagenæ, we may expect to find in our latitudes numerous forms not here recorded.

In arranging the members of this fauna I have adhered to the Linnean method of distinguishing the species by the least fickle features which in every case present themselves. Any other method of systematizing, aiming at an unnatural differentiation of allied forms on the basis of smaller differences, results only in confused crowds of nominations of no distinctive or specific value and of no benefit whatever for our study. Intermediate forms, presenting unimportant differences from the typical species, I therefore range as "varieties" or "allied forms" with or without distinctive denominations.

Any one who for some length of time has been engaged in the study of this class, is familiar with the common occurrence of even great diversity between individuals apparently issued from the same stock. Whenever this is the case we should not be too prone to dissociate such forms under different denominations, the more so as it may be expected that the employed 'characters" are not available for a satisfactory limitation of more validity than what concerns local and individual minor discrepances. Thus, in systematizing this class, it is not at all needed to rank "varieties" as "species", as has been insisted upon by Brady. Although it would be consistent with nature to speak of "variety of variety", we are not compelled to carry out such a scheme, when we coordinate all varieties under their type species. It is neither easier nor more difficult to characterise distinct species amongst an assemblage of forms of this class than is the case within other classes of organisms, particularly when we have to deal with allied forms extending over several geological horizons or areas of the sea. Intermediate forms and varieties are met with at every instance, and this course of things will always bring about a more or less deficient limitation of species and genus, but by giving the notion of species its due compass, and expanding the sphæres of genera, always selecting the more constant differential features, we may overcome a good deal embarassing inconsistences in the system, the chief purpose of which should be to unfold the affinity and origin of the forms.

A discrepance in the features of forms issued from the same stock, is not seldom met with in this class, and this kind of variation has brought about and still causes a great deal of confusion in systematizing. It evidently arises from the various degree of development of the initial segment or primordial embryochamber which is subject to some variation in the same species or even in the same individual or "colony". This variableness seems to concern not only the size but also the power of growth of the embryo-chamber, and results in modifications of the arrangement of the subsequent segments.

If we examine, for instance, a "Flabelline" Frondicularia, Fig. I, *A*, we will readily distinguish a small embryonal or primordial segment (*c*) that brings forth a larval stage with five segments (*l*), disposed in a manner which constitutes the Vaginulina shape or the nearly allied Crepidula form amongst the Nodosarinæ. Subsequent to this stage we find the segments arranged in the peculiar Frondicularia manner, representing the mature or fully developed stadium (*m*), and commencing when the larval stage has completed its fifth segment.

Another modification of the same form is represented in Fig. *B*. The initial segment is here about twice as big as in *A*, or somewhat more, and the larval stage shows but three segments, whereupon the mature Frondicularia stage ensues.

Lastly we meet with the form *C*, generally provided with a larger initial segment, which regarding its power of growth or maturity seems to equalize the whole embryonal and larval condition of the animal, assuming at once the mature Frondicularia segments. In all other respects these three forms resemble each other and ought not be ranked as different species. They have all originated from somewhat differentiated primordial embryo-

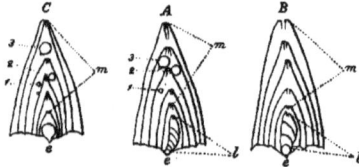

Fig. I. Frondicularia alata D'Orb. with 3 different forms of its larval stage.[1]

segments, and the arrangement of the subsequent segments apparently bears a certain physiological relation to the development of the initial stage.

The least developed primordial segment (*1*) is apparently tending to bring forth a larval stage that inherits the form of the original type (here Vaginulina) and it has to go through a more retarded evolution before it reaches maturity (fig. I *A*). A more powerful initial segment (*2*) gives rise to a more readily performed larval development, its segments being stronger but reduced in number, and more apt to grow into the mature stage (fig. I *B*). The highest developed embryo-segment (*3*) gives rise to a true "Frondicularia" (Fig. *C*) without any intervenient larval stage. Thus it will be obvious that the number and form of the larval segments must become subject to great variation, and stages with two and several segments will therefore be met with in the same species of Frondiculariæ.

Thus, in this instance, we have before us the plainest proceeding of evolution from one type to another, in which an earlier type becomes larva for another type. At last the larval condition is reduced to a single segment and a new form has originated, seemingly standing without much morphological connection with its origin.

[1] The different embryo segments are represented as occupants of the mature segments, but in reality the larger ones are probably generated outside of the chambers.

It must, however, be borne in mind, that the smallest embryo segments not always give rise to a dimorphous colony, but at once assume the mature arrangement of the segments. Such forms are often apt to a pygmean growth. The same mode of evolution in other genera, particularly in Miliolina, has been made the object of most painstaking researches by the experienced Rhizopodologist M. SCHLUMBERGER, who with the utmost skill and sedulity, in numerous papers [1] — mostly during the last seven years published in the Mémoires and Bulletins de la Société Zoologique de France and in Mémoires de la Société géologique de France — has produced masses of examples of this evolutional process in some Miliolinæ. In these excellent expositions it is stated, that the smallest primordial segments give rise to a larval stage generally of *quinqueloculine* arrangement of the segments, often succeeded by a *triloculine* development, whereupon follows the mature biloculine condition.

Instances of a total *triloculine* larval stage will also be met with, being a shorter step to the fully developed *biloculine* stage. The initial segment in such forms is generally somewhat larger than in the last mentioned forms.

As in the instances of polymorphism in *Frondicularia* the differences in the arrangement and number of segments in the larval stage are subject to great variation, so may also the polymorphism in the *Miliolinæ* be looked upon as vacillating gradations of evolution, which cannot afford satisfactory distinctions for establishing species; and that so much the less, when we consider, that initial segments with different power of growth are, as stated above, produced by one and the same individual segment. An arrangement of the larval segments may in several instances appear as being directed by a clear mathematical rule, but which at another time will be eclipsed by irregularities and exceptions. [2]

To show how little attention has been paid to these facts even by reputed Rhizopodologists I will reproduce some of v. REUSS' designations amongst an assemblage of *Nodosarina communis* D'ORB, with varieties, depicted by v. SCHLICHT in his valuable memoir, the Foraminiferen des Septarienthones von Pietzpuhl 1870 (Wien Ak. S. Ber. 62, p. 455).

To any one experienced in dealing with this category of organic forms it will be obvious, that all the forms here (Fig. 11) represented from fig. 1 to fig. 22 must be grouped under *Nodos. communis* D'ORB. var. *consobrina* D'ORB. (a denomination, which will be better substituted by "*pauperata*" D'ORB. the nomen triviale "*consobrina*" being superfluous). As usually the segments of the more developed stages become more or less inflated and present constricted sutures.

Fig. 23—26 represent forms approaching *Nod. Boueana, ovicula* D'ORB. and perhaps the allied *farcimen* SOLD. but may also be considered as a feeble form of the preceding.

Fig. 27—34 are forms of *N. communis* D'ORB. and not distinct from *Nod. Roemeri* and *mucronata* NEUGEB.

---

[1] For the courteous communication of his memoirs I have to acknowledge my great obligations to the author. Valuable informations and suggestions on this subject have also been advanced by VAN DEN BROECK in Bull. R. Soc. Malac. Belge (1893), 28: and in Bull. Soc. Belge Geol. (1893), 7. For transmission of his able papers I stand under great obligation to this author.

[2] For a fuller account of the "dimorphism" see my paper "Om den så kallade verkliga diamorfismen hos Rhizopoda reticulata" 1889, Bih. till K. Sv. Vet. Ak. Handl. 15. 4, No. 2.

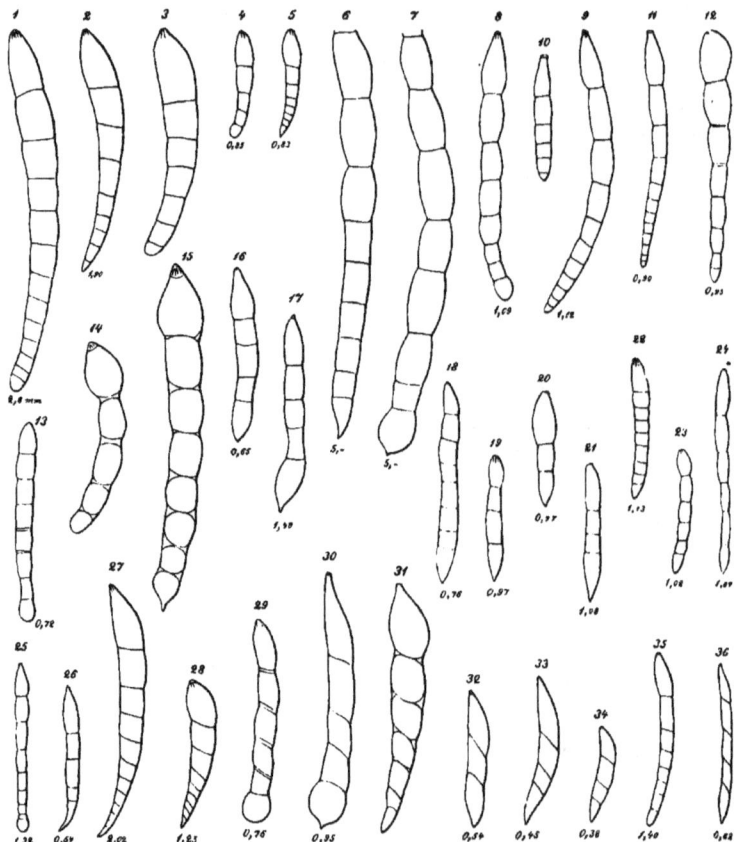

Fig. II. Nodosarina communis D'ORB. with 3 allied forms.

1, 6. 7, 8, 9, 15. 23, 24.   *N. consobrina* (D'ORB.) Rss. — 2. *N. Vernculi* (D'ORB.) Rss. — 3, 4, 10. *N. vermiculum* Rss. — 5. *N. Bötcheri* Rss. — 11. *N. acuticauda* Rss — 12. *N. laxa* Rss. — 13. *N. subæqualis* Rss. — 14. *N. inflexa* Rss. — 16, 18, 20, 21, 26. *N. bicuspidata* Rss. — 17. *N. Henningseni* Rss. — 19. *N. approximata* Rss. — 22. *N. plebeja* Rss. — 25, 29. *N. indifferens* Rss. — 27, 31. *N. inornata* Rss — 28. *N. mucronata* (NEUGEB) Rss. — 30, 35. *N. obliquata* Rss. — 32. *N. abnormis* Rss. — 36. *N. communis* (D'ORB.) Rss. — 33, 34. *N. pygmæa* Rss.

Fig. 35—36 *Nod. communis* D'ORB. approaching the slender form called *badenensis* D'ORB.

Thus, amongst the whole assemblage of figured forms which v. REUSS has distinguished as 19 species, we are unable to discern more than one type species with its 2 or 3 varieties. Many more instances of a thoroughly arbitrary differentiation of forms could be produced, and the chosen one is not to be considered as being of a more extravagant character than that displayed in works of several other writers.

Amongst features and morphological structures which cannot reasonably be considered as of any but relative value for distinctive purposes are to be pointed out the following:

the *absence or presence* of a *marginal wing* or *keel*, of *marginal spines*, of *pseudopodal tubes*, of *"beads"* scattered over or covering the surface, and of *limbation of the sutures*.

Striation of the shell-surface is also a rather unsteady feature, one part of the shell often being striated, another part smooth.

The *number of ridges* is subject to great variation, for instance among the Lagenæ, Nodosarinæ, and Miliolinæ.

The shape of the aperture of some genera exhibits a remarkable degree of variation, with great diversity between the young and older stages in one and the same "colony", particularly amongst Miliolinæ.

The limitation of the genera in this class has been subject to just as many incongruities as the ranging of the species. Several families, particularly Nodosarinæ, Buliminæ, Miliolinæ, offer ample illustrations of the shortcoming of an uncalled for endeavour to dissociate allied types. When the boundary-lines of distinction drawn up between the genera at every instance not hold good, we must fall in with a spurious differentiation encumbering the system with distinctions totally in want of reasonable qualifications. The usual result of such a device is that the student soon will be confronted with forms which reasonably would be ranked under *one* species, but still, according to the artificial differentiation, are disjoined under two—three spurious genera. Such instances are not seldom met with in D'ORBIGNY's and other distinguished authors' arrangement of genera. In the agglutinating groups the institution of genera seems founded on a singularly arbitrarious basis. Nay, the whole order *"arenaceous"* is scarcely based on sound discrimination.

The inconsistency in instituting the *"arenaceous"* Foraminifera as a separate group is shown among such genera as *Textularia, Miliolina* and *Bulimina*, which often exhibit forms with agglutinating test, and still must be ranged with their calcareous congeners. Certain facts seem also to support the view advanced by MELCH. NEUMAYR[1] and others, that a great deal of the homogeneous calcareous Foraminifera have taken their origin from arenaceous forms. It would, therefore, not be surprising to meet with Polystomellæ, Nonioninæ, and other calcareous forms, in the same agglutinating condition. Further in-

---

[1] M. NEUMAYR: Die natürlichen Verwandtschaftverhältnisse der schalentragenden Foraminiferen; Wien, Ak. Sitz. Ber. (1887), 95, 1.

quiries may bring forth several instances of this category, and it would consequently be found that a great deal of arenaceous forms are too nearly allied to the calcareous ones to be grouped so far asunder from one another as usually they have been. NEUMAYR's disposition of the Foraminifera in 5 families, some containing arenaceous genera coordinated with the homogen-calcareous ones, is very acceptable as being more consistent with our notion of the close alliance between the agglutinating and not agglutinating forms of *Textulariæ* and *Miliolinæ*. Whether the different genera in NEUMAYR's system in every instance have been ranged under their proper tribus or place of affinity, is a question that may be subject to a variance of opinions, but his device is worthy of the Rhizopodologists great attention. For deciding in these matters we are in demand of far more extensive assemblages of forms both recent and extinct than as yet have come to notice.

As to the "arenaceous" group, in its present usually adopted arrangement, there would be some reasonable ground for objections even concerning the disposal of its sub-families and their genera. Some of these sub-families are apparently in want of true natural distinctions, for we are not able to consider the nature of the cementing and the smoothness of the surface as of sufficient distinctive value for instituting families and genera (Cnf. BRADY, Challeng. Rep. 9, 1884, p. 63—67). On such fickle ground the tribus Rhabdamminineæ has been splitted up in 9 genera. The Lituolineæ have been subject to the same treatment. The limits between the Nodosarina-like Lituolæ and the Nautiloid ones are as faint as between the true Nodosarinæ and their allied Cristellariæ.

Under the subfamily *Trochammineæ* the most heterogenous forms have been ranged on account of their tests being more finely agglutinated and compacter than in *Lituolineæ;* and far apart from both these subfamilies have been placed the closely allied *Cyclammineæ* on no other ground than the cellular or cancellated condition of the shell-wall in the larger forms of the family, while these features are sometimes wanting in the smaller ones.

With these few remarks on the usually adopted plan of systematizing this class I have intended only to convey to the interested student an idea of the several blanks in our information regarding the real affinities between the genera and forms in question.

For much valuable assistance in preparing this synopsis I have to acknowledge my great obligation to Professor S. LOVÉN who has put at my disposal for examination the whole collection of Rhizopodes belonging to the Swedish State-Museum.

Professor G. LINDSTRÖM, of the Palæozoic Departement of the same Museum, has with unremitting interest kindly communicated to me papers of later dates on the subject which, without such an able assistance probably would have escaped my notice. To Mr. CHARLES DAVID SHERBORN in London I am greatly indebted for communicating to me his most valuable Bibliography of the Foraminifera which publication is indispensable to the study of this class.

Sincere thanks are also due to the well-known Rhizopodologists MM. JOSEPH WRIGHT of Belfast and FORTESCUE MILLETT in Cornwall which gentlemen with the greatest liberality have furnished me with type-samples from their own and other english writers' descriptive papers.

To the Geheimrath CARL MÖBIUS in Berlin I am under obligation for lending me for examination good many examples belonging to the German State Museum; as also to Professor W. DAMES, curator of the Geological-palæozoic Departement of the Museum, for his kind assistance in enabling me to examine the tertiarian and chalk Foraminifera under his care.

To my friend Cav. LUIGI DI ROVASENDA at Sciolze it is a pleasant duty to express my hearty thanks for his efficient assistance and kind hospitality while on a stay in Piemonte last spring in search of tertiarian Foraminifera. I have also to acknowledge the communicating of valuable papers and advices in connection with this essay from Prof. FED. SACCO, Rev. ERM. DERVIEUX, Dr. G. DE AMICIS in Turin and from Dr. C. FORNASINI in Bologna.

ASTRORHIZA SANDAHL.

A. limicola SANDAHL.

Tab. I, figg. 1—3.

Depressa, lenticularis, brachiis 7—15 radiantibus, marginalibus, apice in brachiola irregulariter plerumque divisis; testa minus crassa, nunc maxima e parte e limo nunc e sabulo et glarea constructa; nigrogrisea aut brunnea.

Brachiola flexilia, membranacea, limo et arena sparse intexta, fragilia, caduca. Interdum tumidiuscula.

Fig. 1: forma luxurians e limo et glarea constructa; ad insulas Koster Bahusiæ, profund. metr. 130 (Dr. A. WIRÉN).

Figg. 2—3: formæ solitæ, e fretis insularum Koster; profund. metr. 15—30. (Fig. 3: A. arenaria auctorum).

Astrorh. limicola SANDAHL, 1857: Nya former af Rhizopoder; Sv. Vet. Ak. Öfvers. 14, p. 301, t. 3, ff. 5—6.
Astrodiscus arenaceus E. SCHULZE, 1874, 2te Jahresber. Commiss. Wissensch. Untersuch. deutschen Meere 1872—73, 5, p. 113, t. 2, f. 10.
Hæckelina gigantea BESSELS, 1874: Hæckelina gigantea, ein Protist d. Monothalamien, Jenaische Zeitschr. 9, p. 265, t. 14.
Arenistella elegans, FISCHER et FOLIN, 1875, Les fonds de la mer 2, pp. 26 & 52; FISCHER, 1875, Note sur un type particulier des Rhizopodes (Astrorhiza), Journ. Zool. 4, 1875, p. 510, t. 1, ff. 1—4.
Astrorh. limicola BRADY, 1884, Chall. Rep. 9, p. 231, t. 19, ff. 1—4.

Hab. ad oras Scandinaviæ occidentales, metr. 8—60 passim.

A. arenaria NORMAN.

Tab. II, figg. 4—10.

Depressa aut ventricosa, irregulariter radiata aut ramosa, sæpe alcicorniformis, brachiis apice nunc obtuso nunc extenuato, aperturis singulis.

Testa ex sabulo et arena constructa, brachiis interdum compressis. A præcedente non sat limitanda; notæ a BRADY (Challeng. Rep. p. 232) exhibitæ vagæ, non specificæ, nec materia structuræ notas veras reddit.

Longit. mm. 6—18.

Figg. 4—5: e mari Spetsbergensi; profund. metr. 500—950.

Fig. 6: e fretis insularum Koster Bahusiæ, profund. metr. 15—25.

Figg. 7—8: e sinu Skagerack; profund. metr. 250; Fig. 8: facies marginalis.

Fig. 9: e sinu Söderfjord Norvegiæ; profund. metr. 444 (Dr. C. AURIVILLIUS).

Fig. 10: e sinu Österfjord Norvegiæ; profund. metr. 530 (Rev. NORMAN).

Astrorh. arenaria NORMAN, 1876, "Valorous" Cruise in Davis strait 1875, Proc. Roy. Soc. 25, p. 213.
»   »   BRADY, 1879, Retic Rhizop. Chall. Exped., Qv. Journ. Micr. Sc. (n. s.) 19, p. 43.
»   »   CARPENTER, 1876. On genus Astrorh. described as Hæckelina by Dr. BESSELS; Quart. Journ. Microsc. Sc. (n. s.) 16, p. 221, t. 19, ff. 1—13 (ex parte?).
»   »   BRADY, 1884, Challeng. Rep. 9, p. 232, t. 19, ff. 5—10.

Hab. mare germanicum, sinus norvegicos, mare arcticum; metr. 250—4,200 passim.

A. crassatina BRADY.

Tab. II, figg. 11—15.

Ovalis aut irregulariter clavata, interdum fusiformis aut cylindrica; canali interna plerumque utrinque aperta, sæpe inæquali, passim dilatata; apertura sæpe coarctata, subcirc* ulari; testa crassa ex sabulo fragmentisque testarum et spiculis spongiarum constructa; grisea, rudis; extremum alterum apertura interdum destitutum.

Figg. 11—12: lagenæformis, e mari Spetsbergensi; profund. metr. 4,630; fig. 12: apertura unica.

Figg. 13—15: magis cylindrica, ex eodem loco, profund. metr. 2,670; fig. 14: apertura superna; fig. 15: apertura bulbi initialis.

Astrorh. crassatina, BRADY, 1881, Ret. Rhizop. Chall. Exped., Qu. Journ. micr. Sc. (n. s.) 21, p. 47.
  »   BRADY, 1884, Chall. Rep. 9, p. 233, t. 20, ff. 1—9.

Hab. mare arcticum Spetsbergense metr. 2,670 (CUYDENIUS); metr. 4,630 (v. OTTER); long. mm. 8—10.

STORTHOSPHÆRA E. SCHULZE.

S. albida E. SCHULZE.

Globosa aut subglobosa, rugosa spinisque crassis, brevibus, obtusis, sæpe irregulariter compressis, curvatis et subdigitatis obtecta; aperturis obsoletis.

Plerumque albidogrisea aut fusco-brunnea.
Diam. mm. 2—3.

S. albida SCHULZE, 1875. 2te. Jahresber. Commiss. wissenschaft. Untersuch. der deutsch. Meere 5, p. 113, t. 2, f. 9.
  »   BRADY, 1884, Challeng. Rep. 9, p. 241, t. 25, ff. 15—17.

Hab. in sinubus norvegicis, profund. metr. 320—640 passim (Rev. NORMAN, Dr. APPELLÖF).

SACCAMMINA M. SARS.

S. sphærica M. SARS.

Tab. III, figg. 16—18.

Globosa aut subpyriformis, interdum fusiformis; apertura circularis sæpe in tubulo brevi patens.

Nunc ex arena tenui nunc sabulo structa, plerumque grisea aut ferruginea; magnitudine valde varians, mm. 0.5—3.0.

Interdum bi-triloculata.

Figg. 16—17: exemplum mediocre e fretis insularum Koster Bahusiæ, profund. metr. 50—100.

Fig. 18: pygmæa e sinu Skagerack profund. metr. 580.

Saccammina sphærica M. Sars, 1868, Vidensk. Selsk. Forh. 1868, p. 248.
>   »   Brady, 1884, Challeng. Rep. 9, p. 253, t. 18, ff. 11—17.

Hab. ad oras occidentales Scandinaviæ profund. metr. 2—1,000 sæpe frequentissima; mare Groenlandicum metr. 500 (Prof. Dr. Lindahl, Dr. Öberg),

PSAMMOSPHÆRA E. Schulze.
P. fusca E. Schulze.
Tab. III, fig. 19.

Sphærica aut subsphærica, apertura communis nulla; ex sabulo plerumque crasso constructa; nunc libera nunc affixa.

Forsan a præcedente immerito disjuncta.

Fig. 19: e mari Norvegico profund. metr. 530.

Psammosphæra fusca F. E. Schulze, 1874, 2te. Jahresber. Commiss. Untersuch. deutsch. Meere 1872—73, 5, p. 113, t. 2, f. 8.
>   Brady, 1879, Ret. Rhizop. Chall. Exped., Qu. Journ. micr. sc. (n. s.) 19, p. 27, t. 4, ff. 1—2.
Haeusler, 1883, Astrorh. u. Lituol. Bimammat. zone (Jura); Leonh. u. Bronns Jhb. 1883, p. 57, t. 3, f. 1.
>   Brady, 1884, Chall. Rep. 9, p. 249, t. 18, ff. 1—8.

Hab. ad oras Scandinaviæ occidentales metr. 40—200 cum præcedente minus frequens; Diam. mm. 0.30—3.

TECHNITELLA Norman.
T. legumen Norman.
(Tab. III, figg. 20—27).

Teres aut subcompressa, elongata plerumque paullum curvata; parte primordiali obtusa, rotundata aut apiculata; parte orali plerumque attenuata, apertura irregulariter rotundata interdum valvulata aut cruciata, subproboscidea; albida aut cinerea sæpissime ex spiculis spongiarum contexta, sublævis, flexilis; interdum ex sabulo constructa.

Figg. 20—22: exempla magnitudine varia;
Figg. 23—25: aperturæ diversæ.
Fig. 26: longitudinaliter, fig. 27 transverse secta.

Technit. legumen Norman, 1878, Genus Haliphysema, A. M. N. H. (5) 1, p. 279, t. 16, ff. 3—4.
>   »   Brady, 1884, Chall. Rep. 9, p. 246, t. 25, ff. 8—12.

Hab. ad Koster insulas profund. metr. 100—178 passim. Long. mm. 2.25—4.

CRITHIONINA Goës.

Labyrinthica sive cavernosa aut cavo indiviso centrali pariete subcavernoso prædita; aperturis sparsis aut indistinctis.

C. granum n. s.

Tab. III, figg. 28—33.

Sublenticularis aut subglobosa aut oblonga, plerumque obsolete polygonata, sublævis; aperturis minutis sparsis, interdum nonnullis paullo majoribus unum in locum congregatis; toto interno quam maxime irregulariter cavernoso; e detrito tenuissimo spiculisque spongiarum plerumque structa, friabilis, argillaceo-cinerea.

Fig. 28: exempli lenticularis faciem lateralem exhibens; diam. mm. 2.00.

Fig. 29: facies marginalis ejusdem.

Fig. 30: oblonga, irregulariter obtuse trigona; longit. mm. 4.00.

Fig. 31: exemplum idem de extremo visum.

Fig. 32: facies alius apicalis aperturas aut scrobiculos nonnullos præbens.

Fig. 33: sectio longitudinalis ejusdem, cavernosam structuram exhibens.

Hab. in sinu Skagerack profund. metr. 300 haud frequens (Doct. J. LINDAHL).

C. mamilla n. s.

Tab. III, figg. 34—36.

Plerumque affixa, subglobosa aut semiglobosa, sublævis; pariete crasso, subspongioso; camera centrali subsphærica diametro crassitudinem parietis æquante, indivisa; aperturis minimis sparsis; e detrito tenui structa; argillacea.

Fig. 34: facies superna.

Fig. 35: sectio transversa, canaliculum divisum exhibens.

Fig. 36: facies lateralis.

Hab. in fretis insularum Koster, Babusiæ, profund. metr. 106, in Zostera mortua affixa, haud frequens (Goës) Diam. mm. 1.50.

HALIPHYSEMA BOWERBANK.

H. Tumanoviczii BOWERB. var abyssicola n.

Tab. III, figg. 37—38.

Tubi tenues simplices orificium apicale versus sensim dilatati, e lamina basali, irregulari, nodosa et lobata quasi stolones exeuntes; apertura elliptica, corona spiculorum spongiarum circumdata; ex spinulis spongiarum longitudinaliter dispositis maxime contexta; cinerea; mm. 4.0—5.0 alta.

A forma typica BOWERBANKII statura magis elata et trunco extenuato diversa; typicam etiam nostris in maribus occurrere probabile est.

Fig. 37: truncus.

Fig. 38: apex cum apertura.

Hab. mare Germanicum Lophoheliæ adhærens, profund. metr. 540, haud frequens.

BATHYSIPHON M. Sars.

B. filiformis M. Sars.

Tab. III, figg. 89—41.

Tubus æqualis aut subæqualis, plerumque paullum arcuatus, utrinque apertus, sublævis, segmentatione spuria, internodiis cylindricis, subfusiformibus aut subclavatis; intus indivisus.

Albidus aut griseus ex spiculis spongiarum magna ex parte constructus; subflexilis, tenax.

Fig. 39: forma tenuis.

Fig. 40: » elata.

Fig. 41: longitudinaliter sectus.

Bathysiphon filiformis (M. Sars) G. O. Sars, 1871, Vidensk. Selsk. Forh. 1871, p. 251.
>   >   Brady, 1884, Chall. Rep. 9, p. 248, t. 26, ff. 15—20.

Hab. in sinu Hardanger profund. metr. 350 (Doct. C. Aurivillius); Korsfjord profund. metr. 300 (Norman) haud infrequens; long. mm. 15—25—50; diameter tubi valde varians.

HYPERAMMINA Brady.

H. subnodosa Brady.

Tab. III, figg. 42—54.

Cylindrica, elongata, interdum subclavata, suturis constrictis passim instructa; parte primordiali ovali aut inflate subpyriformi, cujus cavum internum globosum aut ovatum; septis spuriis incompletis aut nullis; apertura lumen tubi totum occupante, aut paullum coarctata. Plerumque grisea-argillacea aut albida, e sabulo rudis.

Emaciatæ, magis attenuatæ, septis subcompletis, interdum occurrunt, Reophaci noduloso approximantes.

Suturis interdum obsoletis, testa magis æquali a Hyperammina friabili difficile distincta.

Fig. 42: exemplum camera primordiali valde inflata.

Fig. 43: apex cum apertura coarctata; Fig. 44: camera primordialis exempli alius; e Atlantico boreali profund. metr. 1,750.

Fig. 45: exempl. e sinu Groenlandico Claushavn profund. metr. 500.

Figg. 46—47, 50—53: emaciatæ in Reophacem vergens, e Groenlandico mari profund. metr. 500.

Figg. 48—49: ex Atlantico boreali profund. metr. 1,750; septis subcompletis.

Fig. 54: in Hyperam. friabilem Brady vergens, e sinu Baffini profund. metr. 1,200.

Hyperammina subnodosa Brady, 1884, Challeng. Rep. 9, p. 259, t. 23, ff. 11—14.

Hab. in sinubus Groenlandiæ, Spetsbergensibus, marique Atlantico boreali profund. metr. 30—2,000; mm. 10—20 longæ; emaciatæ mm. 5—10.

H. elongata BRADY.

Tab. IV, figg. 55—58.

Cylindrica aut subcylindrica, recta aut paullum curvata, nunc lævis polita, nunc sabulo rudis, parte primordiali ovali aut subsphærica, plus minusve inflata; apertura nunc coarctata, nunc lumen tubi totum occupante. Fulva, grisea aut albida.

Forma sabulosa, angustata, curvata, bulbo primordiali sæpe obsoleto a Jaculella difficiliter distinguenda.

Fig. 55: lævis, fulva aut fulvobrunnea, e fretis insularum Koster profund. metr. 80—120.

Fig. 56: sublævis e mari Spetsbergensi, profund. metr. 950.

Fig. 57—58: var. magis sabulosa, curvata, albogrisea, e mari Groenlandico profund. metr. 500, et e fretis Koster profund. metr. 70—180.

Hyper. elongata BRADY, 1878, Ret. & Rad. Rhizop. Arct. Exped. 1875—76; A. M. N. H. (5) 1, p. 433
         (ex parte).
    ›   1884, Chall. Rep. 9, p. 257, t. 23, ff. 4, 7—10.
    v. lœvigata WRIGHT, 1891, For. S. West. off Ireland, Procced. Roy. Irish. Acad. (3) 1,
         p. 466, t. 20, f. 1.

Hab. ad oras Scandinaviæ occidental. metr. 50—170 passim nec non mare Atlant. boreale metr. 1,700; maria Spetsbergensia & Groenlandica metr. 350—890; long. mm. 5—10.

H. friabilis BRADY.

Tab. IV, fig. 59.

Clavata, elongata, aut subcylindrica, testa crassa ex sabulo haud firme coherente et aliquando ex spiculis spongiarum constructa.

A præcedentibus non sat distincta.

Fig. 59: exemplum non sat typicum in H. elongatam vergens, e mari Bahusiæ prope insulam Hållö.

Hyperammina friabilis BRADY, 1884, Chall. Rep. 9, p. 258, t. 23, ff. 1—3, 5—6.
Hyper. elongata BRADY, 1878, Retic. & Rad. Rhizop. Arct. Exp. 1875—76, A. M. N. H. (5) 1, p. 433,
         t. 20, f. 2.
    ›   Goës, 1882, Ret. Rhizop. Carib. Sea, Sv. Vet. Ak. Handl. 19, 4, t. 12, ff. 426—428.

Hab. mare Atlant. boreale metr. 500 rara; ad insulam Hållö Bahusiæ, profund. metr. 60.

H. vagans BRADY.

Tab. IV, fig. 60.

Affixa; tubus æqualis, tortuosus et flexuosus, repens, e bulbo primordiali exiens; plerumque lævis aut sublævis. Rufobrunnea aut fulva; Hyp. elongatæ propinqua.

Fig. 60: exemplum incompletum, ex Atlantico boreali.

Hyperammina vagans BRADY, 1879, Ret. Rhizop. Chall. Exped., Qu. Journ. micr. Sc. (n. s.) 19, p. 33, t. 5, f. 3.
>   »   BRADY, 1884, Chall. Rep. 9, p. 260, t. 24, ff. 1—9.

Hab. Marc Atlant. boreale metr. 1,750 (LINDAHL).

H. ramosa BRADY.

Tab. IV, figg. 61—62.

Tubus irregulariter curvatus et ramosus e bulbo primordiali exiens; truncus primus crassus, rami secundarii sæpe angustiores; testa tennis sabulo et spiculis spongiarum sæpe rudis; albida aut flavida.

Fig. 61: stadium incipiens.

Fig. 62: pars ramosa trunci, e mari Atlantico boreali.

Hyperammina ramosa BRADY, 1879, Rhizop. Chall. Exped., Qu. Journ. micr. Sc. (n. s.) 19, p. 33, t. 3, ff. 14—15.
>   »   BRADY, 1884, Chall. Report. 9, p. 261, t. 23, ff. 15—19.

Hab. mare Atlanticum boreale metr. 1,700 (LINDAHL.) rara; in sinu Hardanger fjord Norvegiæ metr. 350 (Rev. NORMAN).

H. arborescens NORMAN.

Tab. IV, figg. 63—64.

Tubus sublævis tenuissimus, setaceus, dichotomus, erectus, parte primordiali bulbosa affixa, apertura ramorum apicali irregulariter coarctata; nigrobrunnea, purpurescens; bulbo primordiali interdum caret. A Rhizammina algæformi BRADY vix generice distincta. Algæ pusillæ similis; e chitino et sabulo tenui constructa.

Fig. 63: ex Atlantico extra Irlandiam profund. metr. 100; a Cel. Dom. Jos. WRIGHT benigne missa.

Fig. 64: trunci pars magis amplificata.

Hyperammina (Psammatodendrum) arborescens (NORMAN) BRADY, 1881, For. Austr. Hungar. Northpol. Exp. 1872—74, A. M. N. Hist. (5) 8, p. 404.
>   »   »   BRADY, 1884, Chall. Rep. 9, p. 262, t. 28, ff. 12—13, et xylographia, p. 263.
>   »   »   WRIGHT, 1885, For. Down & Antrim; Proceed Belf. nat. Field Club, 1884—85, Append. t. 26, f. 1.

Hab. ad oras Groenlandiæ metr. 36 (NORMAN), altitudine mm. 3—6.

RHABDAMMINA BRADY.

R. linearis BRADY.

Tab. IV, figg. 65—66.

Tubus subcylindricus medio in cameram ovalem aut globosam dilatatus, rectus aut arcuatus, sæpe irregulariter subnodosus; aperturæ terminales plerumque haud coarctatæ;

paries tubi crassus, parte inflata magis tenuis; pallida aut fusca; Rhabdamminæ abyssorum SARS forsan forma pygmæa.

Nostra typo BRADYI validior.

Rhabd. linearis BRADY, 1879, Ret. Rhiz. Chall. Exp., Qu. Journ. Micr. Sc. (n. s.) 19, p. 37, t. 3, ff. 10—11.
> » BRADY, 1884, Chall. Rep. 9, p. 269, t. 22, ff. 1—6.

Hab. mare Atlant. boreale metr. 1670 raru; ad Groenlandiam metr. 100, long. mm. 7—10.

R. abyssorum SARS.

Tab. IV, figg. 67, 68.

Tubi 3—5 radiatim dispositi, nunc tenues, nunc incrassati, plerumque ex camera irregulari centrali exeuntes, cylindrici aut apicem versus attenuati, singuli apertura apicali paullum coarctata præditi; rufobrunnea, scabra.

Fig. 67: brachiis tenuibus, e mari Lofotensi Norvegiæ, profund. metr. 530 (M. SARS).

Fig. 68: magis robusta e sinu Baffini, Groenlandiæ, profund. metr. 400.

Rhabdammina abyssorum M. SARS, 1868, Vidensk. Selsk. Forh. 1868, p. 248 (sine descriptione).
> » BRADY, 1884, Chall. Rep. 9, p. 266, t. 21, ff. 1—13.

Hab. Mare Atlanticum boreale, sinus norvegicos et groenlandicos metr. 350—1,000 passim, locis quibusdam frequentissima; long. mm. usque ad 23.

forma affinis:

Tubo unico subrecto suturis subconstrictis, colore magis fusco.

R. discreta BRADY, 1884, Chall. Rep. 9, p. 268, t. 22, ff. 7—10.
Rhabdopleura DAWSON, 1870, Gulf. St. Lawrence; Canadian Natur. (n. s.) 5, p. 176, f. 7.

Hab. c. præcedente minus frequens.

BOTELLINA CARPENTER.

B. labyrinthica BRADY.

Tab. IV, figg. 69—86.

Tubiformis, apertura foramine irregulari simplici aut poris nonnullis exhibita irregularibus; interne costis et granis sabulosis irregulariter cancellata; parte primordiali plerumque paullum inflata, interne sublævi, pariete tenui, interdum perforata; ex sabulo crasso constructa, sæpe inæqualis, squalida, nido vermis similis.

Forma nostra a typica statura minori, cancellis internis minus evolutis, incompletis differt; grisea aut ferruginea.

Fig. 69, 73: exempla ordinaria.

Fig. 70: tubus longitudinaliter apertus, cancellas exhibens.

Fig. 71: sectio transversa tubi.

Fig. 72: sectio transversa bulbi.
Fig. 74: facies oralis aperturas poriformes præbens.
Figg. 75—80: exempla emaciata cum aperturis apicalibus cujusque diversis.
Figg. 81—86: exempla bulbo reducto perforato.
Figg. 82, 85: aperturæ apicales.
Figg. 83, 86: aperturæ bulbi.

Botellina labyrinthica BRADY, 1881, Retic. Rhizop. Chall. Exp., Qu. Journ. Microsc. Sc. (n. s.) 21, p. 48.
>   >   BRADY, 1884, Chall. Rep. 9, p. 279, t. 29, ff. 8—18.

Hab. ad Koster insulas Sveciæ metr. 35 passim. (LJUNGMAN, GOÉS); long. mm. 10- 15.

JACULELLA BRADY.

J. obtusa BRADY.

Tab. IV, figg. 87—89; Tab. V, figg. 90 -91.

Tubus subulatus angustus, elongatus, rectus aut leniter curvatus, haud segmentatus, parte primordiali (sæpissime abrupta) bulbum minutum irregularem exhibente; apertura tubi plerumque haud constricta;

testa tenuis, firma, plus minusve sabulo rudis, interdum bifurcata; nunc brunnea nunc griseo-albida, verme nematodea sæpe occupata.

Fig. 87: exemplum ordinarium.
Fig. 88: apertura ejusdem.
Fig. 89: exemplum bifurcatum; ambo e freto insularum Koster.
Fig. 90: pygmæa cum facie sua aperturali, e Atlantico boreali profund. metr. 750, (dubia).
Fig. 91: cum facie aperturæ, arena rudissima, e fretis insularum Koster, profund. metr. 100.

Jaculella obtusa BRADY, 1884, Challeng. Rep. 9, p. 256, t. 22, ff. 19—22.

Hab. ad Insulas Koster Sveciæ, metr. 60—180, haud infrequens; long. mm. 10- 15.

HAPLOPIIRAGMIUM v. REUSS.

H. canariense D'ORB.

Tab. V, figg. 92—101.

Discoideum, nautiloideum aut subnautiloideum, subcompressum, utrinque plerumque umbilicatum, anfractu ultimo præcedentes fere toto amplectente, segmentis paullum inflatis numero 7—10—11; suturis impressis, interdum paullum undatis; apertura rima arcuata, marginalis suturalis, interdum labiata; ferrugineum aut albidum, nunc scaberrimum nunc sublæve; interdum pygmæum testa tenui; aliquando valde incrassatum sequenti formæ approximans.

Genus *Haplophragmium* a *Lituola* immerito a REUSS disjunctum, illud *simplicibus* lueve *subdivisis* cameris.

Figg. 92—94: crassum, e sinu Skagerack, profund. metr. 300.
Figg. 95—96: e mari Carico Sibiriæ, profund. metr. 75.
Figg. 97—98: medium, tenue, e Gullmaren sinu Bahusiæ profund. metr. 130.
Figg. 99—101: pygmæum, e mari Spetsbergensi, profund. metr. 400.

Nonionina canariensis D'ORB., 1839, Foram. Canaries, p. 128, t. 2, ff. 33—34.
Placopsilina canariensis PARK & JONES, 1857, For. Const. Norway, An. M. Nat. Hist. (2), 19, p. 301, t. 10, ff. 13—14.
Non. Jeffreysii WILLIAMS, 1858, Rec. For. Gr. Brit., p. 34, ff. 72—73 (minor, lævior).
Lituola nautiloides var. canariensis PARK & JONES, 1865, North. Atl. & Arct. Oc., Phil. Transact. 155, p. 406, t. 15, f. 45.
Haplophragm. rotundidorsatum v. HANTK., 1875, Cläv. Szab. Sch., p. 12, t. 1, f. 2.
»   canariense BRADY, 1884, Chall. Rep. 9, p. 310, t. 35, ff. 1—3, 5.

forma affinis: valde incrassatum.

H. crassimargo NORMAN, 1892, Mus. Normannum, p. 17.
BRADY, Chall. Rep. 9, t. 35, f. 4.

Hab. ad oras Scandinaviæ occidentalis, Groenlandiæ, Spetsbergiæ metr. 10—500 ubique frequens; diam. mm. 2—25. Var. crassimargo præsertim in maribus arcticis.

H. latidorsatum BORNEM.

Tab. V, figg. 102—123.

Nautiloideum, crassum aut subglobosum, plerumque sine umbilico, anfractu ultimo præcedentes amplectente, segmentis numero 4, rare usque ad 8, suturis sæpe minus incisis; fuscum aut ferrugineum; anfractus interdum irregulariter dispositi.

Figg. 102—104: e mari Spetsbergensi, lat. bor. 81° 19', profund. metr. 2,240.
Figg. 105—107: structura magis irregulari, e mari Spetsbergensi, profund. metr. 3,900.
Figg. 108—110: magis compressum, e mari Spetsbergensi.
Fig. 111: faciem marginalem præbens, valde irregulare, in H. turbinatum Br. vergens; ex eodem loco ac præcedens.
Figg. 112—114: minus irregulare, e mari Spetsbergensi profund. metr. 1,030.
Figg. 115—117: fere globosum, e sinu Skagerack, profund. metr. 240.
Figg. 118—120: pygmæum, e mari Spetsbergensi, lat. bor. 72°, profund. metr. 400.
Figg. 121—123: varietas lævior e freto insular. Koster, profund. metr. 180.

Nonionina latidorsata BORNEM., 1855, Septar. Thou. Hermsdorf; Zeitschr. deut. geol. Gesellsch. 7, p. 339, t. 16, f. 4.
Haplophragm. crassum REUSS, 1867, Steinsaltz. Ablag. Wieliczka; Wien Ak. Sitz. Ber. 55, p. 62, t. 1, ff. 1—2.
GOÈS, 1882, Ret. Rhizop. Carib. Sea; Sv. Vet. Ak. Handl. 19, 4, t. 12, ff. 419—420.
Lituola subglobosa M. SARS, 1868, Videnskabs-Sælskab Forbandl., p. 250.
Haplóphragm. latidorsatum BRADY, 1884, Chall. Rep. 9, p. 307, t. 34, ff. 7—10, 14.

Var. minus, sæpe umbilicatum, nitidum, suturis impressis, segmentis turgidis visibilibus numero 4.

Obs. Haplophr. scitulum BRADY, 1884, Chall. Rep., p. 308, t. 34, ff. 11—13, subnautiloideum, minus inflatum, cameris anfractus ultimi 8—11, statura minori forsan in plagis nostris occurrit.

Hab. mare Groenlandicum & Spetsbergense metr. 530 - 4,600 frequentissimum, diam. mm. 1—2; nec non ad oras Scandinaviæ passim metr. 120—350, diam. mm. 0.30—1.5. Var. lævis ad insulas Koster metr. 181.

H. nanum BRADY.

Tab. V, figg. 124 - 127.

Helicoideum, depressum, anfractibus 2—2¹/₂, segmentis anfractus ultimi 5—6—7, suturis paullum incisis, facie spirali subconvexa, facie umbilicali vix-aut subumbilicata, umbilico quasi valvula interdum obtecto; apertura suturalis, marginalis, extra umbilicalis; ferrugineum aut griseo-albidum.

Fig. 124: faciem spiralem,
Fig. 125: faciem oralem,
Fig. 126: faciem marginalem præbens; e mari arctico, profund. metr. 960.
Fig. 127: faciem marginalem, oralem, exempli tumidioris præbens, ex Atlantico boreali, profund. metr. 1,750.

Haplophr. nanum BRADY, 1881, Arct. For. Austr. Hungar. Exped. 1872—74, A. N. Hist. (5) 8, p. 406,
            t. 21, f. 1; Wien. Ak. Dkschr. 43, p. 99, t. 2, f. 1.
    »   »   1884, Challeng. Rep. 9, p. 311, t. 35, ff. 6—8.

Hab. mare Atlant. boreale et arcticum metr. 170—1,750 passim; diam. mm. 0.28.

H. globigeriniforme PARK. & JONES.

Tab. V, figg. 128—133.

Helicoideum, inflatum, anfractibus 2—3; segmentis ultimi anfractus 4—5, globosis; apertura semilunari umbilicali; rufoferrugineum aut albidum, sæpe subpolitum.

Fig. 128: facies spiralis,
Fig. 129: facies oralis,
Fig. 130: facies marginalis exempli e mari Atlant. bor., profund. metr. 1,750.
Fig. 131: facies spiralis,
Fig. 132: facies oralis,
Fig. 133: facies marginalis exempli pygmæi e mari Spetsbergico, lat. boreal. 72°, profund. metr. 390.

Lituola nautiloidea var. globigeriniformis PARK. & JONES, 1865, N. Atl. & Arct. Oc.; Phil. Trans. 155,
            p. 407, t. 15, ff. 46, 47; t. 17, ff. 96, 98.
    »   »   WALLICH, 1862; Northatl. Seabed., t. 6, f. 22.
Lit. globigeriniformis WRIGHT, 1877, Proc. Belf. nat. Field Club. Append. 1876, 1877, t. 4, f. 6.·
? Lituolina irregularis var. globigerinæformis GOËS, 1882, Ret. Rhizop. Carib. Sea; Sv. Vet. Akad. Handl.
            19, 4, p. 141, t. 12, ff. 424—425.
Haplophragm. globigeriniforme BRADY, 1884, Chall. Rep. 9, p. 312, t. 35, ff. 10—11.
    »   »   BALKWILL & MILLET, 1884, For. Galway; Journ. microsc. & nat. Sc. 3,
            t. 1, f. 5.

Hab. mare Atlanticum boreale et arcticum, profund. metr. 300—1,750, passim.

H. glomeratum BRADY.

Tab. V, figg. 134—139.

Nautiloideum, ovale aut subglobosum, plerumque rude, segmentis tumidis 3—4, indistinctis, irregularibus; nunc vix umbilicatum, nunc biconcavum.

Plerumque asymmetricum interdum crassius quam latius; apertura lateralis, suturalis, extra-umbilicalis; ferrugineum.

Fig. 134: facies aboralis,
Fig. 135: facies oralis,
Fig. 136: facies marginalis, e Gullmaren sinu Bahusiæ, profund. metr. 140.
Fig. 137: facies aboralis,
Fig. 138: facies oralis,
Fig. 139: facies marginalis, e mari Groenlandico, profund. metr. 675.

Lituola glomerata BRADY, 1878, Retic. & Radiol. Rhizop. Arct. Exped. 1875—76, A. M. N. H. (5) 1, p. 433, t. 20, f. 1.
Haplophr. glomeratum WRIGTH, 1881, For. S. Donegal, Proc. Belf. Nat. Field. Cl. 1880—81, App., t. 8, f. 1.
   »   »   BRADY, 1882, Österr. Ung. Nordpol. Exped., Wien. Ak. Dkschr. 43, p. 100.
   »   »   BRADY, Challeng. Rep. 9, p. 309, t. 34, ff. 15—18.
   »   »   BALKWILL & MILLETT, 1884, For. Galway; Journ. Micr. & nat. Sc. 3, t. 1, f. 6.

Hab. in Gullmaren sinu Bahusiæ metr. 140 (WIRÉN, CARL AURIVILLIUS); mare arcticum, ad Groenlandiam, metr. 500—600 sat frequens; diam. mm. 0.30—0.40.

H. agglutinans D'ORB.

Tab. V, figg. 140—141.

Lituiforme, stadium juvenile spirale, anfractibus 1—1½, paullum compressum, umbilicatulum, margine rotundato; stadium maturum cylindricum; apertura apicali, circulari aut ovali.

Fig. 140: facies lateralis.
Fig. 141: facies marginalis.

Spirolina agglutinans D'ORB, 1846, Bass. tert. Vienne, p. 137, t. 7, ff. 10—12.
Haplophragm. rectum BRADY, 1876, Carbonif. & Perm. Foram., Pal. Soc. 30, p. 66, t. 8, ff. 8—9.
Spirol. simplex Rss, 1855, Tert. Sch. nördl. & mittl. Deutschl., Wien. Ak. Sitz. Ber. 18, p. 232, t. 2, f. 30.
Haplophr. agglutinans BRADY, 1884, Chall. Rep. 9, p. 301, t. 32, ff. 19—26.

Hab. mare Atlant. boreale metr. 1,704, rarum, (LINDAHL), long. mm. 1.20.

H. pseudospirale WILLIAMS.

Tab. V, figg. 142—151.

Plus minusve distincte lituiforme, complanatum, sæpe late expansum tenuissimum; stadio juvenili sæpe umbilicato, indistincte plano-spirali, septis plus minusve distinctis, apertura angusta elongata apicali; sabulo scabrum.

Ab Haplophragm. foliaceo BRADY vix limitandum.

Figg. 142—147: exempla varia cum faciebus marginalibus cujusque, suturis obsoletis.'
Figg. 148—149: suturis magis distinctis.
Figg. 150—151: tenuissima, Fig. 150 expansum (H. foliaceum BRADY).

Proteonina pseudospiralis WILL., 1858, Rec. For. Gr. Brit., p. 2, t. 1, ff. 2 - 3.
Haplophragm. pseudospirale BRADY, 1884, Chall. Rep. 9, p. 302, t. 33, ff. 1--4.
Haphlophr. foliaceum BRADY, ibid. p. 304, t. 33, ff. 20—25.
     »   »   BALKW. & WRIGHT, 1885, Rec. Dubl. For., Trans. R. Ir. Ac. Sc. 28, p. 330, t. 13, ff. 6—8.
Caf. **Haplophr. calcareum** TERQU., 1886, For. et Ostracod. d'Islande; Bull. Soc. Zool. France, 1886, p. 332,
            t. 11, f. 11.
Nomen. triviale WILLIAMSONI valde ineptum, cum foliaceo BRADYI potius mutandum.

Hab. ad oras Bahusiæ metr 20—170 frequens, long. usque ad 2.90 mm.

H. cassis PARKER.

Tab. V, figg. 152—157.

Marginulinæforme, arcuatum, compressum aut subcompressum, margine externo attenuato; stadio juvenili segmentis semispiralibus paucis; stadio maturo segmentis obliquis, margine interno sive spirali subventricoso; apertura apicali, circulari.

Plerumque griseum, interdum ferrugineum.

Figg. 152, 154, 156: facies lateralis.

Fig. 153: margo spiralis.

Figg. 155, 157: pars summa cum apertura.

Lituola cassis PARK., 1870, For. Gulf. & River St. Lawrence, Canad. Nat. (n. s.) 5, p. 177, 180. f. 3.
Haplophr. cassis BRADY, 1884, Chall. Rep. 9, p. 304, t. 33, ff. 17- 19.

Hab. mare arcticum, Groenlandicum & Spetsbergense, metr. 180—250 minus frequens; long. mm. 2.50.

REOPHAX MONTFORT.

R. scorpiurus MONTFORT.

Tab. V, figg. 158—163; Tab. VI figg. 164—169.

Anguste conicus aut fusiformis, rectus aut curvatus, segmentis subcylindricis plerumque numero 5—6; apertura circulari, interdum reniformi.

Plerumque valde inæqualis et asper, ex sabulo crasso et frustulis testaceis constructus.
Interdum in *Haplophragm. cassidem* vergens.

Forma *Proteonina fusiformis* WILLIAMS (Rec. For. Gr. Brit. p. 1 f. 1) est uni-biloculata segmento ultimo magno fusiformi.

Figg. 158—159: ex Atlantico boreali profund. metr. 320.

Figg. 160—161: in formam fusiformem vergens; e mari Groenlandico, profund. metr. 35—215.

Figg. 162—163: ›fusiformis› e sinu Skagerack, profund. metr. 250.

Figg. 164—166: in H. cassidem vergens, e mari Groenlandico, profund. metr. 35—215.

Fig. 167: subfusiformis e sinu Gullmaren Bahusiæ, profund. metr. 140.

Figg. 168—169: magis regulariter structus, var. dentaliniformi Br. approximans, e Skagerack, profund. metr. 107.

Figg. 170—171: e mari Atlantico boreali, profund. metr. 330.

Reoph. scorpiurus Mntfrt, 1808, Conchyliol. systém. 1, p. 330.
Nodosaria scorpionus d'Orb., 1826, Tab. méthod, Ann. Sc. nat. 7, p. 255, No. 40.
Lituola scorpiurus Brady, 1864, Rhizop. Shetland, Trans. Lin. Soc. 24, p. 467, t. 48, f. 5.
    »   **nautiloides,** var. scorpiurus, Park & Jones, 1865, North. Atl. & Arct. Oc., Phil. Transact, p. 407,
                  t. 15, f. 48.
    »   »   Dawson, 1871, For. Lawrence, A. M. N. II. (4) 7, p. 86, f. 4.
Reoph. scorpiurus, helvetica, Häusler, 1883, Jurassic. Astrorhizoid. & Lituolidæ, Qu. journ. geol. Soc. 39,
                p. 27, t. 2, ff. 7—10.
        »   Brady, 1884, Chall., Rep. 9, p. 291, t. 30, ff. 12—17.
Lituolina scorpiurus Goës, 1882, Ret. Rhizop. Carib. Sea, Sv. Vet. Ak. Handl. 19, 4. p. 136, t. 11, f. 406—409.
Reoph. fusiformis Brady, 1884, Chall. Rep. 9, p. 290, t. 30, ff. 7—11.
    »   »   Br. Park. Jones, 1887, Abroblos Bank, Trans. Zool. Soc. Lond. 12, 7, p. 217, t. 41, f. 18.

Hab. ad oras occidentales Scandinaviæ intr 50—350 passim; mare Atlant. boreale et arcticum intr 90—350 sat frequens; long mm. 2.50.

formæ affines:

1. magis regularis, lævior, segmentis 6—8; a typica vix nec a *R. bacillari* Brady, longiori, segmentis brevioribus, numerosis, sat distincta; sæpe tenuis et fragilis.
Figg. 172—173: exemplum e sinu Skagerack profund. metr. 590.
Figg. 174—175: e mari Baltico, extra Gotlandiam insulam profund. metr. 107.

R. dentaliniformis Brady, Tab. VI, figg. 172—175.
Reoph. dentaliniformis Brady, 1884, Challeng. Rep. 9, p. 293, t. 30, ff. 21—22.
Lituolina scorpiura Goës, 1881, Sv. Vet. Aknd. Öfversigt 1881, 8, p. 33, fig. xylographica.

Hab. in sinu Skagerack metr. 180—590 rara: Mare Balticum passim (Dr. L. Kolmodin), long. mm. 1.0—2.5.

2. paullum compressus, præterea ut præcedens.

R. arcticus Brady, 1881, Arct. For. Austro-Hung. Exped 1872—74, A. M. N. H. (5) 8, p. 405, t. 21, f. 2; Wien. Ak. Dkschr. 43, p. 99, t. 2, f. 2.

Hab. mare arcticum rare, long. mm. 0.40.

3. suturis valde constrictis, segmentis inflatis, globosis aut ovatis.
Fig. 176: fragmentum e mari Spetsbergensi fundo metr. 1,700.
Fig. 177: facies aperturalis ejusdem.
Fig. 178—179: exemplum cum facie aperturali e freto Koster insularum, profund. metr. 150.
Fig. 180: exemplum integrum ex eodem loco.

R. pilulifer Brady, Tab. VI, figg. 176—180.
Reoph. pilulifera Brady, 1884, Challeng. Rep. 9, p. 292, t. 30, ff. 18—20.

Hab. ad Koster insulas Svecicas metr. 100, rara (Goës) long. mm. 2.10.

4. tubiformis segmentis inflatis, distantibus, Tab. VI, figg. 181—186; fragmentis solum inventis non sat describendus, forsan idem ac Reoph. distans BRADY, Chall. Rep., p. 296, t. 31, ff. 18—22.

Figg. 181, 184: cameræ ruptæ.

Figg. 182, 185: aperturæ supernæ.

Figg. 183, 186: aperturæ inferæ cujusque.

Hab. cum præcedente rarus (GOËS); long. mm. 3.50.

R. nodulosus BRADY.

Tab. VI, figg. 187—191.

Nodosariæformis, elongatus, suturis plus minusve constrictis, segmentis ovatis aut ellipticis sæpe inflatis aut pyriformibus. Nunc rectus nunc curvatus (*R. adunca* BRADY), interdum valde angustatus: nunc rudis nunc lævis.

A R. scorpiuro var. pilulifer non sat limitandus.

Fig. 187: exemplum e mari Atlantico boreali profund. metr. 1,750.

Figg. 188—189: pygmæus.

Fig. 190: apertura; ambo e fretis Koster insularum, profund. metr. 106.

Fig. 191: forma adunca BR. e mari Atlant. boreali, profund. metr. 1,750.

R. nodulosa BRADY, 1879, Retic. Rhizop. Chall. Exp., Qu. Journ. Micr. Sc. (n. s.) 19, p. 52, t. 4, ff. 7—8.

»   »   BRADY, 1884, Chall. Rep. 9, p. 294, t. 31, ff. 1—9.

» adunca BRADY ibid. t. 31, ff. 23—26.

Hab. mare Atlanticum boreale intr 1,750 (LINDAHL), long. mm. 5—20; ad Koster insulas (pygmæus et aduncus), profund. metr. 90—180 (GoËs).

R. guttifer BRADY.

Tab. VI, figg. 192—195.

Tenuissimus, moniliformis, segmentis globosis, pyriformibus aut semiglobosis, nunc juxtapositis cohærentibus nunc distantibus; ferrugineus.

Figg. 192—193: e mari Spetsbergensi.

Fig. 194: unilocularis.

Fig. 195: facies oralis.

Reophax guttifera BRADY, 1881, Ret. Rhizop. Chall. Exp., Qu. Journ. micr. Sc. (n. s.) 21, p. 49.

»   »   BRADY, 1884, Chall. Rep. 9, p. 295, t. 31, ff. 10—15.

Hab. mare Spetsbergense metr. 1,250 haud rarus, long. mm. 0.60—1.

R. difflugiformis BRADY.

Tab. VI, figg. 196—198.

Globosus aut ovatus, lagenæformis, unilocularis, collo cylindrico brevi: plerumque rufobrunneus.

Fig. 196: e mari Atlantico boreali, profund. metr. 500.
Fig. 197: e mari Groenlandico, profund. metr. 450.
Fig. 198: facies oralis.

Reophax difflugiformis BRADY, 1879, Ret. Rhizop. Chall. Exp., Qu. Jouro. mier. Sc. (n. s.) 19, p. 51, t. 4, f. 3.
» » BRADY, 1884, Chall. Rep. 9, p. 289, t. 30, ff. 1—5.

Hab. mare Atlanticum boreale et Groenlandicum, metr. 340—500 passim, long. mm. 0.30—0.80.

R. sabulosus BRADY.

Tab. VI, figg. 199—202.

Fusiformis aut conicus, parietibus crassis, ex arena haud firme constructis, segmentis subcylindricis, collo brevi instructis, suturis paullum aut vix incisis; griseo-albidus, interne præsertim collis segmentorum ferrugineo-brunneus; suturæ sæpe indistinctæ.

A *R. cylindrico* BRADY, segmentis non productis, firmius constructo, non sat distinctus.

Figg. 199, 201: exempla duo e mari Atlantico boreali, profund. metr. 1,750.

Figg. 200, 202: facies oralis cujusque.

Lituolina scorpiurus var. **ammophila** GOËS, 1882, Ret. Rhizop. Carib. Sen; Sv. Vet. Akad. Handl. 19, 4, p. 137, t. 410—414.
Reophax sabulosa BRADY, 1882, For. Farõe Chann., Proc. Roy Soc. Edinb. 11, p. 715.
» **rudis** BRADY, 1881, Ret. Rhizop. Chall. Exp.; Qu. Journ. mier. Sc. (n. s.) 21, p. 49.
» » CARP., 1881, The Microscope, p. 563, fig. a, b.
» **sabulosa** BRADY, 1884, Chall. Rep. 9, p. 298, t. 32, ff. 5—6.

Hab. mare boreale metr. 1,750 rarus (LINDAHL) long. 7—10 mm.

APPENDIX.

R. compressus n.

Tab. VI, figg. 203—210.

Plus minusve rudis, complanatus, apertura plerumque lanceolata.

E frustulis granisque testaceis plerumque constructus, Haplophragm. tenuimargini BRADY se approximans.

Figg. 203, 207: facies lateralis.
Figg. 204, 208: facies marginalis.
Figg. 205, 206, 209: facies oralis.
Fig. 210: longitudinaliter sectus, cameris duabus subdivisis.

Hab. mare Caraibicum profund. metr. 530 rarus (GOËS) long. mm. 3—4.

R. procerus n.

Tab. VIII, figg. 413—417.

Fusiformis aut anguste conicus, productus, suturis plus minusve impressis, apertura semilunari valvulata; ex granulis tenuibus calcareis constructus tamen scaber et rugosus.

Fig. 413: exemplum validum.
Fig. 414: facies oralis, apertura valvulata.
Fig. 415: sectio segmenti embryonalis.
Figg. 416—417: pygmæus cum facie orali.
Dentalinæ fœdissimæ REUSS similis

Lituolina foedissima Goës, 1882, Ref. Rhizop. Carib. Sen, Sv. Vet. Akad. Handl. 19, 4, p. 138, ff. 415—418.
Clav. procera Goës, 1889, Om Dimorfismen Rhizop. reticul., Sv. Vet. Akad Handl. Bihang 15, 4, No. 2, p. 9, t. 2, f. 17.

Hab. mare Caraibicum metr. 500 rarus (Goës), long. mm. 3—9.

PLACOPSILINA d'Orb.

P. bulla BRADY.

Tab. VI, figg. 211—215.

Affixa hemisphærica, ovoidea aut circularis, depressa, interne sæpe septis paucis ex parte divisa, aut parietibus costis irregularibus rudibus institutis, apertura sæpe obsoleta; albida.

Ad genus Placopsilinam a BRADY sine ratione justa relatam huc tamen ad interim etiam locavi. Generi Crithioninæ magis propinqua.

Fig. 211: facies superna.
Fig. 212: a margine secta costas internas præbens, e fretis insularum Koster profund. metr. 140.
Fig. 213: facies lateralis exempli Rhabdamminæ inhærentis.
Fig. 214: sectio transversa.
Fig. 215: sectio longitudinalis, septa spuria pauca præbens; e sinu Skagerack profund. metr. 160.

Plac. bulla BRADY, (1881) 1884, Chall. Rep. 9, p. 315, t. 35, ff. 16- -17.

Hab. sinum Skagerack, profund. metr. 160—530 sæpe ad Rhabdamminam affixa (Prof. Hj. Theél, Dr. BoWALLIUS); ad Koster insulas profund. metr. 18—140, ad Fucum sæpe affixa.

HIPPOCREPINA PARKER.

H. indivisa PARKER.

Tab. VI, figg. 216—217.

Conica aut fusiformis, teres, tenuis, sublævis, unilocularis; apertura apicalis, semilunaris, irregulariter circularis aut rima curvata, plerumque limbata. Testa sæpe obsolete quasi segmentata, brunneo-ferruginea.

Fig. 216: facies lateralis.
Fig. 217: facies oralis, apertura irregulari.

Hippocrepina indivisa PARK., 1870, DAWSON Canad. Naturalist (n. s.) 5, p. 176, fig. 2.
           »   PARK., For. Gulf. St. Lawrence, A. M. N. Hist. (4) 7, p. 86, f. 2.
      »   »   BRADY, 1881, For. Austro-hungar. Exped. 1872—74, A. M. N. II. (5) 8, p. 407,
             t. 21, ff. 3—4.
      »   BRADY, Wien. Ak. Dkschr. 43, p. 100, t. 2, ff. 3—4.
      »   BRADY, 1884, Chall. Rep. 9, p. 325, t. 26, ff. 10—14.

Hab. in sinu Olittenbay, Groenlandiæ metr. 25 rara (ÖBERG), long. mm. 0.80—1.10.

HORMOSINA BRADY.

H. globulifera BRADY.

Tab. VI, figg. 218—219.

Nodosariæformis, lævis, segmentis globosis magnitudine raptim increscentibus, suturis profunde incisis; apertura apicali, limbata aut in collo tubulari patescente, circulari, angusta; flavo-ferruginea.

Collum sæpe abbreviatum; segmenta embryonalia uniloculata sæpe occurrunt. Genus Hormosina a genere Reophax non sat limitatum.

Confer: Reoph. nodulosam BRADY cum Hormos. ovicula BRADY et monili BRADY, Chall. Rep. t. 31, ff. 1—9; t. 39, ff. 7—13.

Figg. 218—219: exempla duo ex Atlantico boreali, profund. metr. 1,750.

Hormosina globulifera BRADY, 1879, Ret. Rhizop. Chall. Exped., Qu. Journ. micr. Sc. (n. s.) 19, p.60, t.4, ff.4—5.
      »   »   BRADY, 1884, Chall. Rep. 9, p. 326, t. 39, ff. 1—6.

Hab. mare Atlanticum boreale metr. 1,750 passim (LINDAHL), long. mm. 3.

forma affinis: tenuis, segmentis pyriformibus aut ovatis, suturis valde constrictis.

H. ovicula BRADY, Tab. VI, figg. 220—221.
Hormos. ovicula BRADY, 1879, Ret. Rhizop. Chall. Exp., Qu. Journ. Micr. Sc. (n. s.) 19, p. 61, t. 4, f. 6.
BRADY, 1884, Chall. Rep. 9, p. 327, t. 39, ff. 7—9.

Fig. 220: facies lateralis.
Fig. 221: facies oralis.

Hab. cum typica (LINDAHL), long. mm. 1.50.

TROCHAMMINA PARK. & JONES.

T. inflata MONTAG.

Tab. VI, figg. 222—224.

Helicoidea, spira sæpe subdepressa, anfractibus $2^1/_2$—3, segmentis anfract. ultimi 5—6, plus minusve subinflatis, suturis incisis; facies oralis umbilicata, paullum concava, apertura lateralis, suturalis, brevis, extra-umbilicalis; dilute ferruginea, centro brunnea.

Differentia inter *Trochamminam* & *Haplophragmium*, sive inter subfamilias Trochammineas et Lituolineas est vilis nec juste fundata.

Fig. 222: facies oralis.

Fig. 223: facies aboralis.

Fig. 224: facies marginalis.

Nautilus inflatus MONTAG, 1808, Test. Brit. Suppl., p. 81, t. 18, f. 3.
Rotalina inflata WILL., 1858, Rec. For. Gr. Brit., p. 50, t. 4, ff. 93—94.
Trochammina inflata CARPENT, Introd. 1862, p. 141, t. 11, f. 5.
    »   »   BRADY, 1884, Chall. Rep. 9, p. 338, t. 41, f. 4.
    »   »   BALKW. & WRIGHT, 1885, Rec. Dublin For.; Roy. Ir. Ac. Sc. Trans. 28, p. 331, t. 13, f. 12 (trochoidea).

Hab. mare Balticum metr. 2 passim (LILLIEBORG), diam. mm. 0.50.

T. nitida BRADY.

Tab. VI, figg. 225—230.

Helicoidea, sublævis, depressa, spira plerumque applanata, anfractibus $2^1/_2$—3, segmentis anfr. ultimi 6—9, suturis subimpressis, margine rotundato; facies oralis umbilicata, apertura ut in præcedente; ferruginea, albida aut subrosea.

A forma emaciata Troch. inflatæ nisi spira applanata segmentisque plerumque pluribus distincta.

Figg. 225, 228: facies aboralis.

Figg. 226, 230: facies marginalis.

Figg. 227, 229: facies oralis.

Trochammina nitida BRADY, 1881, Ret. Rhizop. Chall. Exped., Qu. Journ. Microsc. Sc. (n. s.) 21, p. 52.
    »   »   BRADY, 1884, Chall. Rep. 9, p. 339, t. 41, ff. 5—6.

Hab. in Österfjord Norveg. metr. 180 passim (NORMAN), diam. mm. 0,50; mare Spetsbergense metr. 1,200, diam. 0,30 mm.

T. Robertsoni BRADY.

Tab. VI, figg. 231—234.

Nautiloidea, lævis, subdiscoidea, subsymmetrica, utrinque umbilicata, umbilico facici aboralis interdum obtecto, segmentis 4—6 subinflatis; apertura angusta suturalis marginalis; fulvo-ferruginea.

Fig. 231: facies aboralis, umbilico fere obtecto.

Fig. 232: facies marginalis cum apertura.

Fig. 233: facies oralis.

Fig. 234: sectio submediana.

Troch. Robertsoni BRADY, 1887, Synopsis Brit. Rec. For., Journ. Roy. Micr. Soc. (2) 7, p. 893.
    »   »   WRIGHT, 1891, For. S. West off Ireland; Proceed. Roy. Irish Acad. (3) 1, No. 4, p. 469, t. 20, f. 4

Hab. in sinu Gullmaren metr. 140, passim (AX. WIRÉN & C. AURIVILLIUS), in sinu Bukken Norvegiæ (NORMAN) metr. 350, diam. mm. 0.25—0.30.

T. vesicularis n.

Tab. VI, figg. 235—237.

Trochoidea subglobosa, tenuissima, anfractibus 5—7, segmentis anfr. ultimi circiter 5, subamplectentibus, inflatis; facies oralis haud umbilicata.
Dilute ferruginea aut subflava, subpellucida.
Fig. 235: facies aboralis.
Fig. 236: facies oralis.
Fig. 237: facies marginalis.

Hab. mare Spetsbergense metr. 350 rarissima; diam. mm. 0.40.

AMMODISCUS v. REUSS.

A. incertus D'ORB.

Tab. VI, figg. 238—239.

Plano-spiralis, complanatus margine rotundato, anfractibus tenuissimis numerosis usque ad 20; apertura tubus spiralis patens; ferrugineus.
Fig. 238: facies lateralis.
Fig. 239: facies marginalis.

Operculina incerta D'ORB., 1839, For. Cuba, p. 49, t. 6, ff. 16—17.
Trochammina incerta Goës, 1882, Ret. Rhizop. Carib. Sea, Sv. Vet. Ak. Handl. 19, 4, p. 136, t. 11, ff. 404—405, ubi vide synonymiam.
Ammod. incertus BRADY, 1884, Chall. Rep. 9, p. 330, t. 38, ff. 1—3.

Hab. ad oras Scandinaviæ occidentales metr. 140—200 passim; diam mm. 0.40—1.

A. tenuis BRADY.

Tab. VI, figg. 240—241.

Plano-spiralis, anfractibus amplioribus, adultorum 5—7, pars primordialis magna, inflata; flavo-ferrugineus.
Forsan stadium maturum præcedentis potius habendus.
Fig. 240: facies lateralis.
Fig. 241: facies marginalis.

Ammodisc. tenuis BRADY, 1881, Ret. Rhizop. Chall. Exp., Qu. Journ. micr. Sc. (n. s.) 21, p. 51.
»   »   BRADY, 1884, Chall. Rep. 9, p. 332, t. 38, ff. 4—6.

Hab. ad Koster insulas Sueciæ metr. 90 rarus (LJUNGMAN), diam. mm. 1.70.

CYCLAMMINA Norman.

C. pusilla Brady.

Tab. VI, figg. 242—244.

Nautiloidea, lævis aut sublævis, biconvexa, umbilicata, segmentis 11—16, suturis sæpe subundulatis, margine attenuato, pariete testæ vix cancellato; apertura marginalis, suturalis; brunnea aut ferruginea.

A Cycl. cancellata Norman non nisi statura minore cancellisque testæ obsoletis distincta.

Fig. 242: facies lateralis.
Fig. 243: facies marginalis cum apertura.
Fig. 244: sectio mediana.

Cyclammina pusilla Brady, (1881) 1884, Chall. Rep. 9, p. 353, t. 37, ff. 20—23.

Hab. mare Atlanticum boreale metr. 1,750 (Lindahl), diam. mm. 1,20.

WEBBINA d'Orb.

W. clavata Park. & Jones.

Tab. VI, figg. 245—246.

Lævis, affixa, ovata aut rotundata convexa semitubo reptante continuans; fulvo-brunnea; pars postica apertura interdum prædita.

Fig. 245: exemplum apertura postica instructum, e sinubus Norvegicis profund. metr. 300.

Fig. 246: alia, minor.

Trochammina irregularis clavata Park. & Jones, 1860, For. Chellaston; Qu. Journ. Geol. Soc. 16, p. 304.
    »   »   Carp., Introd. 1862, t. 11, f. 6.
Webbina clavata Brady, 1884, Chall. Rep. 9, p. 349, t. 41, ff. 12—16.
    »   »   Wright, 1891, For. S. West off Ireland; Proceed Roy. Irish. Acad. (3) 1, p. 470, t. 20, ff. 2—3.

Hab. ad oras Scandinaviæ occidentales metr. 90—450 passim; ad Groenlandiam metr. 200; long. mm. 1.20 et ultra.

TEXTULARIA d'Orb.

* Verneuilina d'Orb.

V. polystropha v. Reuss.

Tab. VII, figg. 247—255.

Arenacea, bulimineæformis, ovalis aut fusiformis 3—4 serialis; apertura rima obliqua suturalis aut commaformis extra-suturalis;

Interdum segmentis parietibus irregulariter subdivisis (- labyrinthica).

Magnitudine valde varians; mm. 0.90—1.13; color plerumque ferrugineus aut albidus.
Figg. 247, 249, 250: exempla formis variis.
Figg. 248, 251: faciei orales.
Fig. 252: sectio stadii juvenilis transversa, trilocularis, septis paullum undulatis.
Figg. 253—255: sectiones transversæ, tri-quadriloculares, septis undatis, cameris sublabyrinthicis.

? **Bulimina polystropha** REUSS, 1845, Böhm. Kreide 2, p. 109, t. 24, f. 53.
Bulim. scabra WILLIAMS, 1858, Rec. For. Gr. Brit., p. 65, t. 5, ff. 136—137.
Polymorph. silicea SCHULTZE, 1854, Org. Polythal., p. 61, t. 6, ff. 10 —11.
Text. agglutinans var. **polystropha** PARK & JONES, 1865, N. Atl. & arct. Oc., Phil. Transact. 155, p. 371, t. 15, f. 26.
Vern. polystropha BRADY, 1878, Redic. & Rad. Rhizop, Arct. Exped. 1875—76, A. M. Nat. Hist. (5) 1, p. 436, t. 20, f. 9; 1884, Chall. Rep. 9, p. 386, t. 47, ff. 15—17.

Hab. ad oras Sveciæ (etiam in freto et in mari baltico australi) et Norvegiæ, metr. 40—500 aut frequens; nec non ad Spetsbergiam minus frequens.

Clavulinæformis segmentis ultimis stichostegicis, apertura rima brevi recta aut arcuata, apicali, Bigenerinæ digitatæ D'ORB. se approximans, Tab. VII, figg. 256—261. Cnf. WRIGHT, Foram. Belfast, Proc. Belfast. Nat. Field Club, Append. 1885—86, p. 320, t. 26, f. 2.
Figg. 256, 258, 260: exempla formis variis.
Figg. 257, 259, 261: facies aperturalis cujusque.

Hab. in sinu Gullmaren profund. metr. 10 rara (Dr. WIRÉN).

V. pygmæa (EGGER?) BRADY.
Tab. VII, figg. 262—263.

Subtrigona, ovalis aut angusta subæqualis, segmentis magis regulariter in series dispositis, subinflatis globosis; apertura angustissima, suturali aut juxtasuturali, sæpe obsolete limbata; albida aut flavida, forma nostra minus lævigata.
Fig. 262: exemplum ordinarium.
Fig. 263: facies oralis, apertura indistincta.

? **Bulimina pygmæa** EGGER, 1857, Mioc. Ortenburg; LÆONH. & BRONNS, Jbb. 1857, p. 284, t. 12, ff. 10 -11.
? **Text. agglutinans** var. **Vern. polystropha** PARK & JONES, 1865, North. Atlant. and Arct. Oc.; Philos. Transact. 155, p. 371, t. 15, f. 26.
Verneuil. pygmæa BRADY, 1884, Challeng. Rep. 9, p. 385, t. 4, 7, ff. 4—7.

Hab: in sinubus Groenlandiæ metr. 35—50 passim; longit. mm. 0.45.

V. propinqua BRADY.
Tab. VII, figg. 264—266.

Conica, segmentis inflatis plus minusve regulariter in ordinem ternatum dispositis, interdum affixa; apertura basali, umbilicali, suturali sicuti in Textularia; sæpe vestibulum umbilicale exhibente; segmenta ultima interdum ut in Textularia disposita; nostra ferruginea, sublævis aut scabra.

Fig. 264: facies lateralis.
Fig. 265: facies oralis, segmentis binis ultimis textularioideis.
Fig. 266: facies oralis alius cum vestibulo, triserialis.

BRADY, 1884, Challenger Rep. 9, p. 387, t. 47, ff. 8—14.

Hab. mare Atlant. boreale profund. metr. 1,750, sparsa (Dr. LINDAHL), altitudo mm. 2.

** Gaudryina D'ORB.

G. pupoides D'ORB.
Tab. VII, figg. 267—277.

Teres aut subcompressa, cylindrica aut conica, scabra aut sublævis; apertura marginali, suturali, semilunari angusta, aut vestibulo umbilicali, aut poris interdum paucis exhibita; interdum subbigencrinoidea. Nostra argillacea aut nigrocinerea.

Valde variat forma: nunc cylindrica elongata, nunc conica multo dilatata, ut altitudo ad latitudinem ac 3 ad 1.m sit.

A Gaudryina subrotundata BRADY non distincta.

Gaudr. pupoides auctoris ejusdem, Chall. Rep., p. 378, t. 46, ff. 1—4, a typo D'ORBIGNYI valde differens, ad sequentem melius referatur.

Fig. 267: exemplum e mari Atlantico extra Azores insulas; sæpe suturis jugosis.
Fig. 268: facies oralis.
Fig. 269: sectio stadii immaturi transversa ejusdem.
Fig. 270: pygmæa eodem ex loco.
Figg. 271—272: exemplum e Gullmaren, sinu Bahusiæ.
Figg. 273—276: facies oralis, aperturis forma variis.
Fig. 277: facies oralis, apertura poris nonnullis apicalibus substituta exempli magni conici, subbigenerinoidis; eodem ex loco.

G. pupoides D'ORB., 1840, For. Craie blanche Paris, Mém. Soc. géol. Franc. 4, p. 44, t. 4, ff. 22—24; 1846, For. Bass. tert. Vienne, p. 197, t. 21, ff. 34—36.
        GOËS, 1882, Retic. Rhizop. Carib. Sea, Sv. Vet. Ak. Handl. 19, 4, p. 81, t. 6, ff. 177—178.

Obs. Gaudr. rugosa D'ORB. hic cum G. pupoide collata, quia figura hujus d'Orbignyi non sat perspicua est. Text. pupoides var. conica GOËS, l. c., t. 6, ff. 181, 182 forsan ad Gaudr. rugosam D'ORB. est referenda.

Hab. in sinubus Sueciæ occidentalis profund. metr. 10—60 sat frequens, long. 3 mm.; nec non in Norvegicis; scabra; tropicæ læviores.

G. chilostoma Rss.
Tab. VII, figg. 278—280.

Compressa aut subteres, sublævis, segmentis sæpe plus minusve inflatis; apertura, rima brevi, sæpe juxtasuturali, limbata; albida.

Forma nostra a Gaudr. baccata SCHWAG., 1866, For. Kar. Nikob; Novara Reise, geol. Th. 2, p. 200, t. 4, f. 12 vix distincta.

Gaudr. chilostoma Rss, 1865, deutsch. Sept. Thon; Wien. Ak. Dkschr. 25, p. 120. t. 1, ff. 5—7.
Text. labiata Rss, 1860, Crag. v. Antwerp.; Wien. Ak. Sitz. Ber. 42, p. 362, t. 2, f. 17.
Gaudr. pupoides var. chilostoma BRADY, 1884, Chall. Rep. 9, p. 379, t. 46. ff. 5—6.
Gaudr. pupoides Bn., PARK, JONES, 1887, Abrohlos Bank.. Trans. Zool. Soc. Lond. 12, 7, p. 219, t. 42, ff. 7, 8. 9.

Hab. Atlanticum boreale profund. metr. 1,750 rara (Dr. Jos. LINDAHL); mm. 1 et ultra.

*** Textularia D'ORB.

T. agglutinans D'ORB.
Tab. VII, figg. 281—284, 294—303.

Subcompressa aut subteres, stadio primo saepe compresso, marginato, stadio maturo ovali aut subtereti; segmentis adultorum numero variantibus (7—17 utraque in serie), nunc angustis nunc altioribus.

Nostrates ex sabulo crassiore, tropicae ex arena calcarea tenui constructae, quare discrepantiae nonnihil exstat, suturis interdum jugosis. Color ferrugineus, ater aut albidogriseus. Neque a T. saggittula DEFR. neque a T. conica D'ORB. sat limitanda. Ab auctoribus nominibus multis praedita.

Fig. 281: facies marginalis.

Fig. 282: facies lateralis.

Figg. 283—284: faciei orales exemplorum e mari Bahusiae.

Figg. 294—296: exemplum e mari Atlantio extra Azores insulas; facies oralis poris impressis praeter aperturam instructa.

Figg. 297—299: forma jugosa (conf. Text. sagittula D'ORB., 1844, For. Iles Canaries, p. 138, t. 1, figg. 19—21).

Figg. 300—301: exemplum e mari Norvegico.

Figg. 302—303: forma angusta e Skagerack profund. metr. 250.

Text. agglutinans D'ORB., 1839, For. Cuba, p. 144, t. 1, ff. 17—18; 32—34.
» » PARK & JONES, 1865, N. Atl. & arct. Oc., Phil. Transact. 155, p. 369, t. 15, f. 21, T. gramini D'ORB. propior.
» » GOËS, 1882, Retic. Rhizop. Carib. Sea, Sv. Vet. Ak. Handl. 19, 4, t. 5, ff. 140—143.
» » MÖBIUS, 1880, For. Maurit., p. 93, t. 9, ff. 1—8.
» BRADY, 1884, Chall. Rep. 9, p. 363, t. 43, ff. 1—3.

Hab. in Skagerack profund. metr. 250; in sinubus Norvegicis metr. 90—180 sat frequens (C. AURIVILLIUS, NORMAN), long. mm. 2.

forma affinis:

1. Nunc libera nunc affixa, nunc sublævis nunc magis aspera, apertura nunc textularioidea, nunc poris indistinctis substituta; stadio immaturo primo saepe triseriali; ferruginea.

Inter typicam et T. asperam BRADY medium tenens.

T. intermedia n. Tab. VII, figg. 304—307.

Fig. 304: exemplum affixum.
Fig. 305: facies oralis ejusdem, poris minutis prædita.
Fig. 306: facies oralis alius exempli liberi.
Fig. 307: stadium immaturum transsectum, triseriale.

Hab. in sinubus Norvegicis profund. metr. 500 rara (HJ. THÉEL & C. BOVALLIUS).

T. sagittula DEFR. var. cuneiformis D'ORB.

Tab. VII, figg. 288—290.

Lingulata aut lanceolata nunc dilatata nunc angustior, plus minusve complanata, marginata, segmentis plerumque paullum angustatis; apertura textulariæ communi; nostra sublævis grisea.

T. cuneiformis D'ORB., 1839, For. Cuba, p. 147, t. 1, ff. 37—39.
T. sagittula Goès, 1882, Ret. Rhizop. Carib. Sea, Sv. Vet. Akad. Handl. 19, 4, t. 5, ff. 144—146 (in sequentem transiens); ubi etiam synonymiam vide, p. 75; a qua abstrahenda: Text. corrugata COSTA (forsan Bigenerina Capreolus D'ORB.), Text. folium PARK & JONES; T. variabilis var. spathulata et difformis WILL.. (acc. BALKWILL & WRIGHT Bolivinæ).
  »   »   BRADY, 1884, Chall. Rep. 9, p. 361, t. 42, ff. 17, 18.
        »   BALKWILL & WRIGHT, 1885, Rec. Dublin For., Trans. R. Irish Ac. Sc. 28, p. 332, t. 13, ff. 16, 17, in sequentem transiens.

Hab. ad oras Bahusiæ profund. metr. 50—90 sat frequens, formis sagittula et Williamsoni interse commixtis, long. 1—1,90 mm.

formæ affines:

Magis dilatata, plus minusve cuneata, nunc marginata nunc margine obtuso, nunc scabriuscula nunc sublævis, segmentis plerumque paucioribus, altioribus; grisea; a typica non sat distincta.

T. Williamsoni Goès, Tab. VII, figg. 285—287.

T. cuneiformis WILLIAMS, 1858, Rec. For. Great Brit., p. 75, t. 6, ff. 158—159.
?T. anceps Rss, 1859, Westph. Kreidef., Wien. Ac. S. Ber. 40, p. 234, t. 13, f. 2.
T. gramen BRADY, 1884, Challeng. Rep. 9, p. 365, t. 43, ff. 9· 10.
  »   »   BALKW. & WRIGHT, 1885, Rec. Dublin For., Transact. R. Irish Ac. Sc. 28, p. 332, t. 13, ff. 13—14.

Hab. ad oras Bahusiæ profund. metr. 40—90 passim, alta mm. 1,25.

APPENDIX

Compressa, margine segmentorum in angulum plus minusve producto, segmentis angustis.

T. pectinata v. Reuss, Tab. VII, figg. 291—293.

1849, Neue For. Österreichs; Wien. Ak, Dkschr. 1, p. 381, t. 49, ff. 2—3; 1867, Steinsalz Ablag.
Wieliczka; Wien. Akad. Sitz. Ber. 55, p. 98, t. 3, f. 11.
Obs. T. Mariae D'Orb., 1846, Russ. tert. Vienne, p. 246, t. 14, ff. 29—31 et Plecanium spinulosum
v. Reuss, 1867, Steinsalz. Ablag. Wieliczka; Wien. Ak. Sitz. Ber. 55, p. 65,
t. 1, f. 3 segmentis altioribus et paucioribus ad T. Williamsoni potius
correspondentes.

Hab. tropicis in maribus profund. metr. 300—500. Long. mm. 1.

**** Bigenerina D'Orb.

B. nodosaria D'Orb.

Tab. VII, figg. 313—323.

Arenacea aut calcarea agglutinans, scabra, stadio larvali sive immaturo compresso plane textularioideo, stadio maturo nodosarino; apertura foramen minutum, saepe excentrice locutum; grisea aut albida.

Fig. 313: e mari Germanico.
Fig. 314: facies oralis e, apertura minuta.
Fig. 315: longitudine secta.
Figg. 316, 320, 322: exempla robusta e mari Caraibico; facies laterales.
Figg. 317, 321: facies marginales.
Figg. 318, 319, 323: facies orales.

Bigen. nodosaria D'Orb., 1826, Tab. méth., An. Sc. nat. 7, p. 261, No. 1, t. 11, ff. 9—12, Mod. 57.
» pusilla Roem., 1838, Tert. Meeres. Norddeutschl., Leonh. & Bronns Jhb., 1838, p. 384, t. 3, f. 20.
agglutinans D'Orb., 1846, Russ. tert. Vienne, p. 238, t. 14, ff. 8—10.
T. agglutinans var. nodosaria Park. & Jones, 1865, N. Atl. & Arct. Oc., Philos. transact. 155,
p. 371, t. 15, f. 25; t. 17, f. 80.
Fifs Modell. 12.
Goës, 1882, Ret. Rhizop. Carib. Sen; Sv. Vet. Ak. Handl. 19, 4, t. 5, ff. 133—139.

Obs. Text. angittula forma bigenerina Goës ibid. p. 78, t. 5, ff. 159—160 inter Textulariam
et Clavulinam medium tenet, quare est a Bigener. nodosaria D'Orb. separanda.

Terrigi, 1880, Foram. Vaticani; Atti Accad. Pontif. 33, p. 192, t. 2, f. 28.
Brady, 1884, Chall. Rep. 9, p. 369, t. 44, ff. 14—18.

Hab. mare Germanicum profund. metr. 180; Stoksund sinum norvegicum metr. 150—180 rara (Norman), long.
1.6; ex mari carnibico min. 2.60.

B. digitata d'Orb.

Tab. VII, figg. 324 – 343.

Arenacea, stadio larvali teretiusculo aut subcompresso, irregulariter bi-triseriali; apertura semilunari, valvulari, aut fissura recta, interdum poro rotundato; cameris interdum imperfecte subdivisis; ferruginea aut albida.

Vern. polystrophæ, ex qua verosimiliter originata, magis quam Bigen. nodosariæ et Text. aliis affinis. Nomen triviale in *"digitus"* rectius commutandum. Clavulinæ communi sæpe valde similis.

Fig. 324: exemplum typicum e Gullmaren sinu Bahusiæ.

Figg. 325—331: oriticia formis variis.

Fig. 332: sectio transversa, cameram subdivisam præbens.

Fig. 333: exemplum stadii primi.

Fig. 334: apertura ejusdem.

Figg. 335—336: sectiones transversæ ejusdem IV-serialibus segmentis.

Fig. 337: exemplum tenue, stadio immaturo magis reducto.

Fig. 338: sectio longitudinalis stadii immaturi.

Fig. 339: sectio longitudinalis stadii maturi, segmentorum trabeculis rudimentariis.

Figg. 340—341: sectiones transversæ, stadiorum juvenalium structuram subdivisam exhibentes.

Bigonorina (Gemmulina) digitata d'Orb., 1826, Tab. Méth. An. Sc. nat. 7, p. 262, No. 4, Mod. 58, (cum forma auctorum non sat congruens).

»   Brady, 1864, Rhiz. Shetland., Trans. Lin. Soc. 24, p. 468, t. 48, f. 8.

»   Park & Jones, 1865, For. N. Atl. & Arct. Oc., Phil. Transact. 155, p. 371, t. 17, f. 81.

»   Parker, Jones, Brady, 1865, Nomenclat. Foram., A. M. Nat. Hist. (3) 16, p. 28, t. 2, f. 61.

Brady, 1883, Chall. Rep. 9, p. 370, t. 44, ff. 19—24.

Hab. ad et extra oras Bahusiæ Sueciæ metr. 40—350 sat frequens; nec non in sinubus Norvegiæ usque ad profund. metr. 350; long. 2—2.5 mm.

SPIROPLECTA Ehrenb.

S. biformis Park. & Jones.

Tab. VII, figg. 308 - 312.

Arenacea, compressa, linearis, linguæformis aut subulata, margine obtuso; interdum valde attenuata, angusta, spira 3—4 loculari, sæpe abrupta, etiam obsoleta, Gaudryinæ filiformi Berthelin similis; ferruginea. Spiroplecta prælonga Wright, 1885, Cretac. For. Keady Hill; Proc. Belfast Nat. Field Club, Append. 1885—86, p. 329, t. 27, f. 3, nostræ formæ valde propinqua.

Figg. 308—309: forma tenuis, e Gullmaren sinu Bahusiæ, profund. metr. 124.

Fig. 310: forma latior, eodem e loco.

Fig. 311: e Jacobshavn, Groenlandiæ, profund. metr. 52.

Fig. 312: e Storfjord Spetsbergiæ, profund. metr. 180.

T. **agglutinans** var. **biformis** PARK & JONES, 1865, North. Atl. & Arct. Oc., Philos. Transact. 155, p. 370, t. 15, ff. 23—24.
T. **biformis** BRADY. 1878, Retic. & Radiol. Rhizop. North. Pol. Exped. 1875—1876; A. M. Nat. Hist. (5) 1, p. 436, t. 20, f. 8; 1884; Challeng. Rep. 9, p. 376, t. 45, ff. 25—27.

Hab. In Gullmaren sinu Bahusiæ metr. 150 (C. AURIVILLIUS, A. WIRÉN); extra Helsingborg metr. 30; ad Groenlandiam (OBERG), Spetsbergiam, metr. 50—270 (SMITT & GOES). Long. mm. 0.40—0.70.

VALVULINA D'ORB.

V. conica PARK & JONES.

Tab. VIII, figg. 342—352.

Plus minusve elevate conica, basi orbiculari aut ovali, segmentis amplectentibus, triserialibus; apertura minima, rima suturali aut juxtasuturali, interdum duplicata (rimulis binis plus minusve sepositis); ferruginea, affixa, tapete e granulis arenæ minimis constructa circumdata.

Fig. 342: minuta.
Fig. 343: facies oralis cum rimis duabus.
Fig. 344: magis elata.
Fig. 345: facies oralis ejusdem elliptica, apertura minima.
Fig. 346: minus elata.
Fig. 347: angustior.
Fig. 348: sectio transversa.
Fig. 349: longitudine secta.
Fig. 350: facies oralis ejusdem.
Fig. 351: facies oralis alius, orbicularis, pro apertura rimulis duabus sepositis.
Fig. 352: Facies spiralis.

V. **triangularis** PARK & JONES, 1857, For. Const. of Norw., A. M. Nat. Hist. (2) 19, p. 295, t. 11, ff. 15—16.
var. **conica**, PARK & JONES, 1865, North. Atlant. & Arct. Oc., Philos. Transact. 155, p. 406, t. 15, f. 27.
» BRADY, 1884, Chall. Rep. 9, p. 392, t. 49, ff. 15—16.

Hab. extra oras Sueciæ occidentales, profund. metr. 90—500 passim; in sinubus Norvegiæ profundis sat frequens; mm. 1.20 alta.

V. fusca WILLIAMS.

Tab. VIII, figg. 353—355.

Explanato-convexa aut late conica, ceterum ut præcedens, a qua vix specifice distincta.

Fig. 353: facies spiralis.
Fig. 354: facies oralis.
Fig. 355: facies latero-marginalis.

40 AXEL GOËS, ARCTIC AND SCANDINAVIAN FORAMINIFERA.

Rotalina fusca WILLIAMS, 1858, Rec. Foramf. Gr. Brit., p. 55, t. 5, ff. 114—115.
V. fusca BRADY, 1884, Challeng. Rep. 9, p. 392, t. 49, ff. 13—14.

Hab. in Stoksund sinu Norvegico profund. metr. 150—180 passim (NORMAN); diam. mm. 1.30.

APPENDIX.

CLAVULINA D'ORB.

C. communis D'ORB.

Stadio immaturo teretiusculo aut subcompresso, segmentis plus minusve regulariter triserialibus; segmentis stadii nodosarini aliquando in collum breve productis; apertura plerumque semilunari, valvata; cinerea, aut albida, structura saepe magis calcarea quam silicea.

Genus Clavulinae retineo, quia origo ejus a Valvulina incerta.

Clavulina communis D'ORB., 1826, Tab. méthod., An. Sc. nat. 7, p. 268, No. 4.
» 1846, Bass. tert. Vienne, p. 196, t. 12, ff. 1—2.
Friés mod. 10.
Verneuilina communis PARK & JONES, BRADY, 1866, Crag. For., Palæogr. Soc. 19, t. 3, f. 19.
» » VAN DEN BROECK, 1876, For. Barbadoes, Ann. Soc. Belg. microsc. 2. p. 136, t. 3, f. 14.
Clavulina communis BRADY, 1884, Challeng. Rep. 9, p. 394, t. 48, ff. 1—3.

Hab. mare Atlanticum extra Hispaniam profund. metr. 1,500 (J. LINDAHL).

formae affines:

1. Attenuata, sublævis, stadio immaturo teretiusculo aut paullum compresso, interdum valde reducto saepe biseriali, apertura porata rare valvulata; albida.
   Interdum a Big. digitata D'ORB. vix distinguenda.

C. lævigata GOËS, Tab. VIII, figg. 356—367.

Figg. 356—357: exempla duo e mari Caraibico.
Fig. 358: longitudine secta.
Figg. 359—367: faciei orales.

BRADY, Challeng. Rep. 9, t. 48, ff. 9—12.
GOËS, Retic. Rhizop. Carib. Sea, Sv. Vet. Ak. Handl. 1882, 19, No. 4, t. 5, f. 162—164.

Hab. mare Caraibicum profund. metr. 800 rara. Longitudo mm. 1.50—2.50.

2. Siliceo-arenacea, cylindrica, interdum subconica aut fusiformis brevis, stadio immaturo brevissimo indistincto, apertura valvata aut foramine minuto aut poris sparsis minutissimis instituta, aliquando obsoleta; cameris interdum rudimentis trabecularum obsolete subdivisis; plerumque ferruginea, aspera; suturis interdum obsolete scrobiculatis.

C. cocæna GÜMB., Tab. VIII, figg. 368—377.

Fig. 368: exemplum cylindricum angustum.
Fig. 369—371: aperturæ formis variis.
Fig. 372: sectio transversa stadii nodosarini dissepimentorum rudimentis.
Fig. 373: apertura obsoleta.
Fig. 374: sectio transversa stadii immaturi.
Fig. 375: sectio longitudinalis.
Fig. 376: forma obconica.
Fig. 377: facies oralis ejusdem.

Gümb., 1868, For. Nordalp. Eocän, K. Bay. Wiss. Akad. Abh. 10, 2, p. 601, t. 1, f. 2.
Valvulina triangularis v. **cocæna** Goës, 1881, Sv. Vet. Ak. Handl. 19, 4, p. 88, t. 11, ff. 401—403.

Hab. mare Caraibicum metr. 800 passim (Goës).

C. parisiensis d'Orb.

Tab. VIII, figg. 378—386.

Stadio immaturo pyramidali trigono, apertura sæpe protrusa, plerumque valvulata, semilunari aut poris instituta, præterea ut in præcedente; stadio maturo interdum suturis valde constrictis; grisea.
Fig. 378: exemplum ordinarium, e mari Caraibico.
Figg. 379—382: aperturæ formis variis.
Fig. 383: sectio transversa stadii immaturi.
Fig. 384: valde attenuata var. humili Brady se approximans; e mari Caraibico.
Figg. 385—386: facies orales aperturam præbentes.

Clavulina parisiensis d'Orb., 1826, Tab. Méth., Ann. Sc. Nat. 7, p. 268, No. 3; Mod. 66.
Text. triquetra **Bigenerina** Goës, 1882, Ret. Rhizop. Carib. Sea; Sv. Vet. Ak. Handl. 19, 4, p. 85, t. 6, ff. 185—186.
Cl. parisiensis Brady, 1884, Chall. Rep., p. 395, t. 48, ff. 14—18.

Hab. mare Atlant. latitud. bor. 40°, long. occident. 10°, profund. metr. 300—1,500 haud rara (Lindahl). Mare Caraib. profund. metr. 500 (Goës), spiculis spongiorum interdum immixtis; Long. mm. 3.

forma affinis:

Stadio immaturo biseriali, lanceolato, valde compresso textulario-ideo; apertura semi-lunari, valvulata, aut rimula aut poris duplicatis aut foramine subrotundo instituta; stadium larvale interdum valde abbreviatum; calcarea scabra.
Bigenerinæ nodosariæ d'Orb. simillima. Ubi forma nostra recte locetur, incerti sumus.

C. textularioidea n., Tab. VIII, figg. 387—399.

Fig. 387: exempli magni facies lateralis.
Fig. 388: sectio longitudinalis ejusdem.
Figg. 389—390: facies marginalis exemplorum 2 minus adultorum.
Figg. 391—396: facies orales, aperturis formis variis.
Figg. 397—398: exempla 2, stadio immaturo valde reducto.
Fig. 399: sectio longitudinalis juvenis.   Omnes e mari Carnibico.

Textularia sagittula forma Bigenerina Goës, 1882, Retic. Rhizop. Carib. Sen, Sv. Vet. Ak. Handl. 19, 4, p. 78, t. 5, ff. 159—160.

Hab. mare Caraibicum profund. metr. 300 frequens.   Long. mm. 3—5; interdum pygmæa.

C. Soldanii Park. & Jones.

Tab. VIII, figg. 400—407.

Conica aut ovalis aut cylindrica, interdum valde angustata, stadio immaturo plus minusve obliterato rare distincte evoluto, cameris septis radialibus divisis; apertura cribriformis vel dendritica interdum semilunaris valvata aut rima curvata instituta.

Nunc pulvere tenuissimo nunc granulis crassioribus constructa, plerumque calcarea. Dissepta numero variantia, plerumque 10—16, alternum brevius; interdum pauciora (3—6). Statura et forma valde variantes.

Fig. 400: forma cylindrica, stadio immaturo aperte clavulinæformi insignis.
Fig. 401: sectio longitudinalis ejusdem.
Fig. 402: sectio transversa stadii immaturi.
Fig. 403: sectio transversa stadii maturi.
Fig. 404: facies oralis, apertura insolita.
Fig. 405: forma tenuis rarissima.
Fig. 406: facies oralis apertura simplex valvata.
Fig. 407: sectio transversa segmenti stadii maturi tridivisi.

Lituola Soldanii Park & Jones, 1860, Rhizop. Mediterr., Quart. Journ. geol. Soc. 16, p. 307.
? Nodos. dubia d'Orb., 1826, Tabl. Méth., An. Sc. nat. 7, p. 252, No. 10.
Lituola Soldanii Carpenter, 1862, Introduct., t. 6, fig. 43.
        »   v. intermedia van den Broeck, 1876. For. Barbade; Ann. Soc. Belg. microsc. 2, p. 74, t. 2, ff. 1, 3, 4, 6.
Haplostiche Soldanii Brady, 1884, Chall. Rep., p. 318, t. 32, ff. 12—18.
Valvul. triangularis v. polyphragma Goës, 1882, Ret. Rhizop. Carib. Sen, Sv. Vet. Akad. Handl. 19, 4, p. 87, t. 11, ff. 390—400.

Hab. mare Caraibicum metr. 300—500 vulgaris (Goës), usque ad mm. 7 alta. Ad Insulas Azores metr. 600 (Smitt & Ljungman).

C. rudis COSTA.

Tab. VIII, figg. 408—412.

Cylindrica, ovate-cylindrica aut subglobularis, stadio immaturo obsoleto ad segmenta 1—2 redacto, segmento ultimo in collum breve producto, apertura semilunari valvulata instructo. Nunc sublævis e pulvere tenuissimo agglutinata nunc sabulo scabra; nunc valida mm. 6—8, nunc pygmæa 1 mm. solum longa, lævigata Sagrinæ similis; valvula aperturæ sæpissime indistincta.

Fig. 408: exemplum ovoideum.

Fig. 409: facies oralis ejusdem.

Figg. 410—412: aperturæ forma variæ.

Glandulina rudis COSTA, 1855, Marna terz. Messina, Mem. Nap. 2, p. 142, t. 1, ff. 12—13.
Clavulina cylindrica v. HANTKEN 1875, For. Clávul. Szabói Schicht., Jahrb. ungar. geol. Anstalt 4, p. 18, t. 1, f. 8.
  „ KAHRER 1877, Hochquell. Wasserleit., Abh. geol. Reichsanst. Österreichs. 9, p 373, t. 16, f. 4.
GOËS 1882, Retic. Rhizop. Carib Sea, Sv. Vet. Akad. Handl. 19, 4, t. 4, f. 77—81 (ad Uvigerinam dimorpham PARK & JONES lævitatis causa relata, in fig. 81 est apertura neglecta).
C. cylindrica BRADY 1884, Challeng. Rep. 9, p. 396, t. 48, f. 32—38.
Clav. rudis FORNASINI, 1893, For. Messin., R. Accad. Sc. Instit. Bologna (5) 3, t. 1, ff. 13—14.

Hab. mare Carnibicum metr. 300 vulgaris (GOËS), ad Azores metr. 350 frequens (SMITT, LJUNGMAN).

CASSIDULINA D'ORB.

C. lævigata D'ORB.

Tab. VIII, figg. 418—420.

Lenticularis, plerumque valde compressa et marginata aut anguste carinata, segmentis utrinque 4—5, centro nunc nudo nunc involuto, poris nunc obscuris nunc distinctis.

Cassidulina lævigata D'ORB. 1826, Ann. S. Nat. 7, p. 282, No. 1, t. 15, ff. 4, 5, Mod. 41.
  „ „ WILLIAMS. 1858, Rec. For. Gr. Brit., p. 6, ff. 141—142.
  „ „ PARK. & JONES 1857, For. Coast of Norway., Ann. Mag. Nat. Hist. (2) 19, t. 11, ff. 17 —18; fidem 1865, North. Atl. & Arct. Oc.; Phil. Transact., 155, p. 377, t. 15, ff. 1—4.
  „ „ BRADY 1884, Chall. Rep. 9, p. 428, t. 54, ff. 1—3, t. 15, ff. 1—3.
C. punctata HSS. 1849, Neue For. Österr.; Wien Ak. DkSchr. 1, p. 376, t. 48, f. 4.
C. sicula SEG. 1862, Rhizop. Catan; Accad. Gioenia Atti (2) 18, p. 27, t. 1, f. 7 (separ.)
C. pulchella D'ORB. 1839, For. Amer. merid, p. 57, t. 8, ff. 1—3.
C. lævigata BRADY, PARK., JONES 1887, Abrohlos Bank Trans. zool. Soc. Lond. 12, 7, p. 221, t. 43, f. 11.

Hab. extra litora occidentalia Sveciæ et Norvegiæ profund. metr. 70—180; nec non ad Spetsbergiam et Groenlandiam metr. 180—500, haud rara. Diam. mm. 0.65.

C. crassa D'ORB.

Tab. VIII, figg. 421—422.

Disciformis, margine obtuso rotundato, segmentis plus minusque inflatis utrinque 4—5, centro sæpe involuto; a præcedente non sat distincta.

Fig. 421: facies lateralis.
Fig. 422: facies marginalis.

C. crassa D'ORB. 1839, For. Amér. mérid. 5, p. 56, t. 7, ff. 18—20; 1846, Buss. tert. Vienne, p. 213, t. 21,
    ff. 42—43.
C. obtusa WILL. 1858, Rec. For. Gr. Brit., p. 69, t. 6, ff. 143—144.
C. lævigata v. crassa PARK. & JONES 1865, North Atl. & Aret. Oc., Phil. Transact. 155, p. 377, t. 15,
    ff. 5—7, t. 17, f. 64.
? C. Margareta KARR. 1877, Hochqu. Wasserleit., Abh. geol. Reichsanst. Österr. 9, p. 386, t. 16, f. 52.
C. crassa BRADY 1884, Chall. Rep. 9, p. 429, t. 54, ff. 4—5.

Hab. mare Germanicum profund. metr. 180 rara; pygmæa mm. 0.28.

C. Bradyi NORMAN.

Tab. VIII, figg. 423—426.

Sublituiformis aut ovalis, plus minusve compressa, segmentis primis helicostegicis
aut semi-helicostegicis sicuti in præcedentibus dispositis, ceteris oblique stichostegicis bi-
serialiter interse alternatis.
Fig. 423: facies lateralis.
Fig. 424: facies marginis convexi.
Fig. 425: facies oralis exempli e mari Germanico.
Fig. 426: latior, ex Österfjord Norvegiæ.

(NORMAN M. S.) BRADY 1884, Chall. Rep., p. 431, t. 54, ff. 6—10.
Goës, 1882, Retic. Rhizop. Caribb. Sea, 8v. Kgl. Vet. Akad. Handl. 19, 4, t. 4, figg. 111—113 (magis infuta).

Hab. in mari Germanico metr. 180; Österfjord, Norvegiæ metr. 180--360 rara; long. mm. 0,30—0,42 (NORMAN).

forma affinis:
stenostegica, numero segmentorum plus quam duplo majore, Tab. VIII, fig. 427.

Hab. cum præcedente rara, long. 0.42 mm.

APPENDIX.

EHRENBERGINA v. REUSS.

E. serrata RSS.

Tab. VIII, figg. 428—430.

Subtriangularis aut late ovalis aut rhomboidea, subcompressa, biconvexa; aut æqui-
lateraliter subtrigono pyramidalis; angulo segmentorum externo in spinam producto; segmentis primis nunc subhelicostegicis vel potius faciem oralem ("ventralem") versus inclinatis, nunc testa tota stichostegica.
Formas trigonas a compressis disjungere non proprium est. Variat margine interno
segmentorum spinis instructo, ut facies oralis ("ventralis") series spinarum binas exhibeat.

Fig. 428: facies dorsalis s. lateralis convexa.
Fig. 429: facies "ventralis" aperturam præbens.
Fig. 430: apex cum apertura.

Ehrenbergina serrata Rss., 1840, Neue For. Österr., Wien. Ak. DkSchr. 1, p. 377, t. 48, f. 7.
       - BRADY, 1884, Chall. Rep. 9, p. 434, t. 55, ff. 2—7 (trigona, spinosa).
GOËS, 1882, Retic. Rhizop. Car. Sea, Sv. Vet. Akad. Handl. 19, 4, t. 6, ff. 183—184 (trigona, pro Verneulina spinosa Rss tunc habita).

Hab. extra Azores insulas profund. metr. 400 passim (SMITT, LJUNGMAN). Long. mm. 0.50—0.55.

BULIMINA D'ORB.

B. pyrula D'ORB.

Ovoidea aut ellipsoidea plus minusve inflata, anfractu ultimo præcedentes fere obvelante, ut anfractus 1—2 solum appareant; vitrea, tenuissima; a sequente non specifice limitanda.

Bulimina pyrula D'ORB, 1846, Bass. tert. Vienne, p. 184, t. 11, ff. 9—10 (minuta, inflata).
? **B. caudigera** D'ORB., 1826, Tab. méthod., Ann. Sc. nat. 7, p. 270, No. 16, Mod. No. 68 (magis angustata).
B. auriculata, turgida BALL., 1850, Microsc. exam. of soundings, Smitsonian contributions 2, 3, p. 12, figg. 25—31.
B. pyrula BRADY, 1884, Chall. Rep. 9, p. 399, t. 50, ff. 7—10.

Hab. cum sequente præsertim in maribus arcticis profund. metr. 300—530 minus frequens.

forma affinis: Statura et forma præcedentis nunc lata nunc angustata, segmentis nunc lente nunc raptim increscentibus, quo formæ variæ sint natæ, anfractibus apparentibus 3—4—5.

B. ellipsoides COSTA, Tab. VIII, figg. 431—436.

Figg. 431—434: exempla duo cum facie orali utriusque, e mari Bahusiæ.
Fig. 435: major, subcylindrica, e mari Groenlandico, profund. metr. 340.
Fig. 436: angusta e sinubus Norvegicis, profund. metr. 100—150.

COSTA, 1854: Palæont. Nap. 2, p. 265, t. 15, f. 9.
Bul. pupoides var. GOËS, 1882, Ret. Rhizop. Carib. Sea, Sv. Vet. Akad. Handl. 19, 4, t. 4, ff. 86—87.
Bul. Presli var. **pyrula** PARK. & JONES, 1865, N. Atl. a. Arc. Oc., Philos. Trans. 155, p. 372, t. 15, ff. 8—9 (in præcedentem vergens).
Bul. tenera Rss., 1867, Steinmnlz Ablag. Wieliczka, Wien. Ak. Sitz. Ber., 55, p. 94, t. 4, ff. 11—12.
Bul. ovata BRADY, 1884, Chall. Rep. 9, p. 400, t. 50, f. 13.
Bul. doliolum, ovoides TERQU., 1886, For. & Ostrac. d'Islande; Bull. Soc. zool. France, 1886, pp. 333, 334, t. 11, ff. 17—18, 20.
B. socialis BORNEM., 1855, Septar. Thon Hermsdorf, Zeitschr. deut. geol. Gesellsch. 7, p. 342, t. 16, f. 10.

Ab his non sat distincta:

B. ovata D'ORB, 1846, Bass tert. Vienne, p. 185, t. 11, ff. 13—14.
B. ovata, propinqua STACHE, 1865, Novara Reise, geol. Th. 1, pp. 266, 267, t. 24, ff. 14, 16.
B. affinis D'ORB., 1839, For. Cuba, p. 105, t. 2, ff. 25—26.
B. pupoides var. GOËS, 1882, Ret. Rhizop. Carib. Sea, Sv. Vet. Akad. Handl., 19, 4, t. 4, ff. 84—85.
B. affinis BRADY, 1884, Chall. Rep. 9, p. 400, t. 50, f. 14.
? **B. ovulum** ALTH., 1850, Umgeb. Lemberg., Haid. nat. hist. Abh. 3, p. 264, t. 13, f. 18.

Hab. in sinubus Sveciæ occidentalis, Norvegiæ, Groenlandiæ, profund. metr. 130 - 300 sat frequens. Long. mm. 1—1.80.

B. marginata d'Orb.

Tab. IX, figg. 439—444.

Ovata margine segmentorum infero serrato aut spinoso, anfractibus visibilibus 5—7. Bul. pupoides v. spinulosa Will., 1858, Rec. For. Gr. Brit., p. 62, t. 5, f. 128, et B. aculeata Brady, 1884, Chall. Rep., p. 406, t. 51, ff. 7—9 non sat distingvendæ; formæ intermediæ sæpe occurrunt.

Figg. 439—444: exempla tria cum facie orali cujusque.

Bul. marginata d'Orb., 1826, Tab. méth. Ann. Sc. nat. 7, p. 269, No. 4, t. 12, ff. 10—12.
   „   „   Brady, Park., Jones, 1887, Abrohlos Bank, Trans. zool. Soc. Lond. 12, 7, p. 220, t. 43, ff. 7, 10.
Bul. pupoides v. marginata Will., 1858, Rec. For. Gr. Brit., p. 62, t. 5, ff. 126—127.
Bul. marginata Brady, 1884, Chall. Rep. 9, p. 405, t. 51, ff. 3—5.
Bul. acanthia Costa, 1854, Pal. Nap. 2, p. 335, t. 13, ff. 35—36.

Hab. ad oras Sveciæ occidentales (etiam in freto), Norvegiæ et ad Spetsbergiam profund. metr. 50—270 sat frequens. Long. mm. 0.60—0.80.

B. subteres Brady.

Tab. IX, figg. 445—453.

Elongate aut abbreviate ovalis, anfractibus 2—3 in spiram plus minusve regulariter retortam dispositis, apertura rima umbilicali aut juxta-umbilicali in margine segmenti infero, rare summo segmento propinqua, instituta; lævissima, interdum paullum compressa.

A Robertina arctica d'Orb., 1846, Bass. tert. Vienne, p. 203, t. 21, ff. 37—38 non distingvenda, quare nomen Bul. arctica Bul. "subtereti" sit præferendum.

A Bul. elegantissima d'Orb., 1839, For. Amér. mérid., p. 51, t. 7, ff. 13—14; Williams, 1858, Rec. For. Gr. Brit., p. 64, t. 5, ff. 134—135; Brady, 1884, Chall. Rep., p. 402, t. 50, ff. 20—22, nisi anfractu hujus ultimo multo longiore segmentisque angustioribus et situ aperturæ medio segmento ultimo distincta. Inter formam utramque medium tenet B. elegantissima Goës, 1882, Ret. Rhiz. Carib. Sea, Sv. Vet. Akad. Handl. 19, 4, p. 66, t. 4, ff. 88, 89, 95—98.

Bulimina imperatrix Karr., 1868, Mioc. Kostej., Wien. Ak. S.Ber. 58, p. 176, t. 4, f. 11, est inter B. pupoidem et B. subteretem notis ambarum prædita.

Figg. 445, 447: exemplum paullum compressum e mari Spetsbergico.
Fig. 446: summum segmentum ultimum ejusdem.
Figg. 448—449: e mari Norvegico.
Figg. 450—452: e. apertura subapicali e mari Groenlandico.
Fig. 453: subcompressa (= Robertina arctica d'Orb.) e mari Spetsbergico.

Bul. subteres BRADY, 1881, Österr. Ung. Nordpol. Exped., Wien. Ak. DkSchr. 43, p. 101; Qv. Journ. Microsc. Sc. 21 (n. s.), p. 55; WRIGHT, 1880, Rec. For. South. Donegal, Proc. Belf. Nat. field Club, Append. 1880--81, t. 8, ff. 2, 2 a.
» » BRADY, 1884, Chall. Rep. 9, p. 403, t. 50, ff. 17—18.
B. elegantissima BRADY, 1878, Retic. & Rad. Rhiz. Northpol. Exp., A. M. N. (5) 1, p. 436, t. 21, f. 12.
B. Presli v. **elegantissima** PARK. & JONES, 1865, N. Atl. & Arct. Oc., Philos. Transact. 155, p. 374, t. 15, ff. 12—17.
B. Parkeri TERQU., 1886, For. & Ost. d'Islande, Bull. Soc. Zool. France 1886, p. 334, t. 11, f. 19.

Hab. mare Spetsbergiæ, Groenlandiæ, Norvegiæ metr. 60—350 passim, nec non ad Bahusiam rara. Long. mm. 0.50—0.80.

forma affinis:

late ovoidea, abreviata, subinflata, anfractibus 2—3, apertura ut in typica, sæpe altera extramimbilicali suturali; forsan megasphærica typi.

Forma a Robertina austriaca v. REUSS (1849, Neue For. Österr. Wien. Ak. DkSchr. 1, p. 375, t. 47, f. 15), paullum discrepans, structura tamen huic propinqua. B. brevi D'ORB. (1839, For. Craie bl. Paris; Mém. Soc. géol. France 4, p. 41, t. 4. ff. 13—14) etiam affinis.

B. Normani n. Tab. IX, figg. 437—438.

Hab. in sinubus Norvegicis profund. metr. 100—150 rara (NORMAN). Long. mm. 0.50—0.65.

B. convoluta WILLIAMS.

Ovate auricularis, compressa aut subcompressa, segmentis primis subhelicoideis, ceteris irregulariter bi-triserialibus, suturis sæpe limbatis.

Cassidulinis, præsertim C. Bradyi NORMAN quam Buliminis magis propinqua.

B. pupoides var. **convoluta** WILLIAMS, 1858, Rec. For. Gr. Brit., p. 63, t. 5, ff. 132—133 (fig. vix correcta).
B. convoluta BRADY, 1884, Chall. Rep. 9. p. 409, t. 113, f. 6.

Hab. in sinubus Norvegicis ut Stoksund et extra Sartoroe metr. 70—200 paucas collegit Rev. NORMAN; pygmæn, long. 0.50—0.60 mm.

** Virgulina.

V. squamosa D'ORB.
Tab. IX, figg. 454—456, 460.

Producta, anguste lanceolata, compressa aut subcompressa, segmentis in series binas regulariter dispositis, nunc productis angustis, septisque obliqvatis, nunc paullo brevioribus, septisque magis transversis, sæpissime non inflatis.

In luce transeunte sæpe quasi sabulosa, nebulosa.

Figg. 454—455: exempla e mari Caraibico, var. Schreibersianæ approximantia; a: facies apicalis segmenti ultimi.

Figg. 456, 460: e sinubus Norvegiæ; a: summum segmentum utrumque ultimum.

Virgulina squamosa D'ORB., 1826, Ann. Sc. nat. 7, p. 267, No. 1, Mod. 64; ROEMER, 1838, Nordd. Meeressand., LEONH. & BRONNS, Jhb. 1838, p. 386, f. 39
Virg. punctata D'ORB., 1839, For. Cuba, p. 139, t. 1, ff. 35—36.
Bul. compressa BAILEY, 1850, Microscop. exam. of soundings, Smithson. Contribut. 2, art. 3, p. 12, ff. 35—37.

? **Virg. tegulata** v. Reuss., 1846, Böhm. Kreide 1, p. 40, t. 13, f. 81.
Bul. Presli var. (**Virg.**) **squamosa** Park. & Jones, 1865, North. Atl. & Arct. Or., Phil. Transact. 155, p. 375,
    t. 15, f. 19.
Bul. squamosa Goës, 1882, Retic. Rhizop. Carib. Sea; Sv. Vet. Akad. Handl. 19, 4, p. 67, t. 4, ff. 99
    —100, 106—107.
Virg. squamosa. Brady, 1884, Chall. Rep. 9, p. 415, t. 52, f. 9.

Obs.   Boliv. textilarioides v. Rss. (1862, Norrd. Hils. & Gault., Wien. Aknd. Sitz Ber. 46, p. 81, t. 10, f. 1) et
    Boliv. elongata v. Karr. (1878, For. Luzon, Bolet. Map. geol. Españ. 7, 2, p. 23, t. F, f. 8) non
    sat distincte, nisi magis complanatae paulloque latiores.

Hab. in sinubus Norvegicis profund. metr. 180—530 passim. Long. mm. 0.60—1.15.

formae propinquae:

1. septis plus minusve transversis, segmentis brevioribus, plus minusve inflatis ut
    margo paullum qvasi lobatus saepe videatur; poris paucis sparsis majoribus
    praecipue secundum suturas congregatis.
    A Boliv. (Textularia) laevigata Williams, Rec. For. Gr. Brit., p. 77, f. 168,
    segmentis inflatis distincta. Haec Bol. tegulatae Rss. 1850, Kreidemerg. Lem-
    berg, Haid. Abh. 4, 1, p. 45, t. 4, f. 12 valde propinqua.
    A Boliv. textilarioides Rss. suturis magis transversis discrepans, nec non seg-
    mentis inflatis; secundum Cel. Dom. Balkwitt & Wright (Roy. Ir. Ac. Sc.
    28, p. 334) eadem ac Bol. laevigata Williams; illa "textilarioides" tamen
    suturis magis obliquatis, segmentis plus productis ovalibusque Virg. squa-
    mosae d'Orb. magis propinqua.

V. obscura Goës, Tab. IX, figg. 457—458.

Fig. 457—458 exempla e mari Germanico profund. metr. 180; a: summum segmen-
    tum ultimum utrumque cum apertura.

Boliv. textilarioides Brady, 1884, Chall. Rep. 9, p. 419. t. 52, ff. 23?, 24—25, ab hoc non vere distincta.
Virg. texturata Brady, ibid., p. 415, t. 52, f. 6.
B. squamosa var. Goës, 1882, Ret. Rhizop. Carib. Sea, Sv. Vet. Akad. Handl. 19, 4, t. 4, ff. 105, 108.

Hab. sinus Norvegicos profund. metr. 500 (Revnd. Norman), mare Germanicum profund. metr. 180 rara,
    pygmaea mm. 0.45 longa.

2. segmentis plus minusve irregulariter dispositis, structura Buliminis propriis se
    proximante; praeterea ut in typica. Juvenes Buliminis veris saepe similes,
    forsan ut "Bul. ovata" d'Orb. interdum relatae.

V. Schreibersiana Czjzek, Tab. IX, figg. 459, 461—472.

Fig. 459: exemplum minutum e mari Norvegico.
Fig. 461: subteres, e Gullmaren Bahusiae.
Fig. 462: Bul. affini d'Orb. similis, juvenis, a: summum segmentum cum aper-
    tura, ex eodem loco.
Fig. 463: magis angusta, ex eodem loco.

Fig. 464> Buliminæ veræ similis, e mari Norvegico.

Fig. 465: e Gullmaren sinu; apertura dilatata; *a:* facies apicalis.

Figg. 466—67: exemplum a latere utroque visum; *a:* facies apicalis.

Fig. 468: exemplum minus.

Figg. 469—470: facies lateralis et marginalis alius, omnia e mari Spetsbergico.

Figg. 471—472: facies lateralis et marginalis alius, e Gullmaren Bahusiæ.

Virgul. Schreibersiana CZJZEK, 1847, For. Wien. Beck., Haid. nat. Abh. 2, p. 147, t. 13, ff. 18—21.
Bul. marginata *β* **attenuata** PARK. & JONES, For. Coast Norw., A. M. N. Hist. (2) 19, t. 11, f. 35.
Bul. Presli v. (Virg.) **Schreibersii** PARK. & JONES, 1865, N. Atl. & Arct. Oc.; Philos. Trans. 155, p. 375, tt. 15, 17, ff. 18, 72 —73.
Bul. pupoides v. **compressa** WILL., 1858, Rec. For. Gr. Brit., p. 63, t. 5, f. 131.
Virg. schreibersiana BRADY, 1884, Chall. Rep. 9, p. 414, t. 52, ff. 1—3.
Bul. squamosa var. GOËS, 1882, Ret. Rhiz. Carib. Sea, Sv. Vet. Akad. Handl. 19, 4, p. 67, t. 4, ff. 101, 106.
Virg. affinis Schreibersi SCHWAG, 1883, Libysche Wüste, Palcontographica 30, p. 112, t. 4, f. 11.

Obs. Bul. pupoides var. fusiformis WILLIAMS. Rec. For. Gr. Brit., p. 63, t. 5, f. 129 ad var. Schreibersianam forsan referendo; specimina a Dom. Jos. WRIGHT e mari Irlandico benigne communicata huic qvidem valde propinqun.

Hab. ad oras Sveciæ, Norvegiæ, Spetsbergiæ profund. metr. 40—890 haud infrequens; long. mm. 0.84 et ultra.

3. brevior, crassior, sæpe subovalis, segmentis plus minusve inflatis, primis in seriem quasi semihelicoideam sæpe dispositis, ceteris nunc magis oblique nunc magis transverse dispositis, testa in luce transeunte sæpe nebulosa.

V. subsquamosa EGGER, Tab. IX, figg. 473—474.

Fig. 473: facies marginalis.

Fig. 474: facies lateralis concava.

EGGER, 1857, Mioc. Ortenb., LEONH. & BRONNS Jhb. 1857, p. 295, t. 12, f. 19—21.
Virg. tenuis SEG., 1862, Rhizop. Catania., Accad. Gioen. Atti (2) 18, (Srp.) p. 26, t. 2, f. 2.
BRADY, 1884, Chall. Rep. 9, p. 415, tab. 52, ff. 7, 8, 11.

Hab. in sinu Gullmaren profund. metr. 150 rara (AURIVILLIUS & WIRÉN); minuta mm. 0.80 longa.

*** Bolivina.

B. punctata D'ORB.

Tab. IX, figg. 475—480.

Lanceolata aut sublinearis, plerumque compressa, regulariter textularioidea. Poris nunc obscuris, nunc distinctis; a B. squamosa nisi segmentis angustioribus, crebrioribus et poratione plerumque magis exquisita non distincta; interdum minus compressa.

Fig. 475: angusta Virgulinæformis e sinu Codano; *a:* facies apicalis.

Figg. 476—77: latiores, e mari Atlantico extra Azores insulas, profund. metr. 540.

Fig. 478: in Virg. squamosam vergens, e mari Bahusiæ; poratione obsoleta; *a:* facies apicalis.

Fig. 480: forma eadem, e mari Spetsbergensi; *a:* facies apicalis.

Bolivina punctata D'ORB., 1839, Voy. Am. mér. For., p. 61, t. 8, ff. 10—12.
Boliv. antiqua D'ORB., 1846, Bass. tert. Vienne, p. 240, t. 14, ff. 11—13.
   »   »   EGGER (ex parte), 1857, Mioc. Ortenb., LEONH. & BRONNS, Jhb. 1857, p. 294, t. 12, f. 26.
Boliv. catanensis SEG., 1862, Rhizop. foss. Catania, Accnd Gioenia Atti (2) 18 (Sep.), p. 29, t. 2, f. 3.
Boliv. punctata BRADY, 1864, Rhizop. Shetl., Trans. Lin. Soc. 24, p. 468, t. 48, f. 9.
? **Boliv. elongata** v. HANTK., 1875, Clávul. Szaboi Sch., Mittheil. Jhb. K. Ungar. geol. Anstalt 4, p. 65, t. 7, f. 14.
Boliv. punctata MöB., 1880, For. Maurit., p. 94, t. 9, ff. 9--10.
   »   »   BRADY, 1884, Chall. Rep. 9, p. 417, t. 52, ff. 18—19.

Hab. ad oras occidentales Sveciæ, Norvegiæ nec non Spetsbergiæ profund. metr. 150—270 passim; long. 0.30—0.90 mm.

formæ affines:

1. paullum dilatata, costis tenuibus sparsis longitudinalibus paucis:

B. ænariensis COSTA.

Brizalina ænariensis COSTA, 1854, Palæont. Nap. 2, p. 297, t. 15, f. 1.
Bul. punctata GOËS, 1882, Ret. Rhizop. Carib. Sea, Sv. Vet. Akad. Handl. 19, 4, t. 4, ff. 114--115.
Boliv. ænariensis BRADY, 1884, Chall. Rep. 9, p. 423, t. 53, ff. 10—11.
   »   »   BRADY, PARK., JONES, 1887, Abrohlos Bank, Transact. Zool. Soc. 12, 7, p. 221, t. 43, ff. 2, 4, 5.

2. dilatata, subtriangularis, anceps lanceolata, segmentis plerumque angustioribus.

B. dilatata RSS, Tab. IX, figg. 482—86.

Fig. 482: exemplum e sinu Bergensium Norvegiæ.
Fig. 483: facies oralis ejusdem.
Fig. 484: e Rasvaag Norvegiæ.
Fig. 485: facies marginalis ejusdem.
Fig. 486: e sinu Skagerack.

v. REUSS, 1849, Neue For. Österr., Wien. Ak. DkSchr. 1, p. 381, t. 48, f. 15.
Text. variabilis v. **spathulata** WILL., 1858, Rec. For. Gr. Brit., p. 76, t. 6, ff. 164—165.
GOËS, 1862, Ret. Rhizop. Carib. Sea, Sv. Vet. Akad. Handl. 19, 4, t. 4, ff. 124—126.
BRADY, 1884, Chall. Rep. 9, p. 418, t. 52, ff. 20—21.

Obs. Boliv. robusta BRADY, 1884, Chall. Rep., p. 421, t. 53, ff. 7—9 ab hac notis specificis non sat distincta. Vide descriptionem auctoris, ubi nota characteristica nulla est relata. Specimina originalia a Rev. NORMAN benigne tradita nullum specificum exhibent.

Hab. cum forma typica passim.

B. difformis WILLIAMS.

Triangularis compressa, dilatata, segmentis paullum inflatis angulo externo in acumen producto, quare margo serratus.

Textularia variabilis var. **difformis** WILLIAMS, 1858, Rec. For. Gr. Brit., p. 77, t. 6, ff. 166—167.
Bolivina pygmœa BRADY, (1881) 1884, Chall. Rep. 9, p. 421, t. 53, ff. 5, 6.

Hab. in sinu Österfjord Norvegiæ, profund. metr. 178—360 (NORMAN). Long. mm. 0.25—0.30.

B. plicata (D'Orb.) Brady.

Tab. IX. figg. 487—488.

Modice compressa, suturis irregulariter sinuato-impressis, impolita, scabra quasi arenacea, in luce transeunte obsolete areolata aut undato-irregulariter interrupte subcostata; poris non sat distinctis; linea suturali mediana plerumque impressa, apertura lamina perforata aut lingva interdum semiocclusa.

Fig. 487: facies lateralis, cum impressionibus suturalibus.

Fig. 488: facies oralis.

D'Orb., 1839, For. Amér. mérid., p. 62, t. 8, ff. 4--7.
Brady, 1870, For. Brack. Water., An. Mag. Nat. H. (4) 6, p. 302, t. 12, f. 7.

Hab. in Bukkefjord Norvegiæ profund. metr. 180—260 (Norman). Long. 0.42.

UVIGERINA D'Orb.

U. pygmæa D'Orb.

Tab. IX, figg. 496—501.

Ovalis aut subcylindrica, teretiuscula, segmentis triserialibus plus minusve inflatis omnibus aut aliqvot costatis aut lineatis; interdum fere lævis costis valde obsoletis, a U. canariensi D'Orb. haud multum diversa.

Figg. 496—97: cylindrica, U. tenuistrata v. Reuss.

Figg. 498—501: typicæ.

U. pygmæa D'Orb., 1826, Tab. méthod, An. Sc. nat. 7, p. 269, t. 12, ff. 8, 9, Mod. 67.
,   ,   Goës, 1882, Ret. Rhizop Carib. Sea, Sv. Vet. Akad. Handl. 19, 4, p 59, t. 4, ff. 68—70
             (ubi vide Synonymiam).
,   ,   Brady, 1884, Chall. Rep. 9, p. 575, t. 74, ff. 11—14.
,   ,   Brady, Park., Jones, 1887, Abrohlos Bank, Trans. Zool. Soc. Lond. 12, 7, t. 45, f. 1.

Obs. Uv. tenuistriata Rss., 1870, For. Pietzpuhl, Wien. Ak. Sitz. Ber. 62, p. 485; v. Schlicht Foramf.
Pietzpuhl, t. 22, ff. 34—36; Brady, 1884, Chall. Rep., p. 574, t. 74, ff. 4—7; cylindrica, elongata,
striata, non sat distincta.

Hab. ad oras occidentales Scandinaviæ profund. metr. 20--90 sat frequens. Long. 1 mm. circiter.

U. angulosa Will.

Tab. IX, figg. 502—509.

Triquetra plus minusve costata aut lævis; a varietate triquetra U. canariensis difficile distingvenda.

Figg. 502—507: læves c. faciebus oralibus.

Fig. 508: costulata.

Fig. 509: facies oralis ejusdem.

U. angulosa WILLIAMS, 1858, Rec. For. Gr. Brit., p. 67, t. 5, f. 140.
> » BRADY, 1884, Chall. Rep. 9, p. 576, t. 74, ff. 15—18.
Synonymiam vide GOÊS, 1882, Ret. Rhizop. Carib. Sea; Sv. Vet. Akad. Handl. 19, No. 4, p. 60.

Hab. ad oras occidentales Scandinaviæ præsertim Norvegicis profund. metr. 20—530 sat frequens nec non ad
Spetsbergiam et Groenlandiam passim profund. metr. usque ad 1,600; minor, long. 0.50—0.70 mm.

* Sagrina.

S. dimorpha PARK. & JONES.

Tab. IX, figg. 510—511.

Subcompressa, sublinearis, suturis sæpe undulate impressis aut foveis instructis,
collo aperturæ brevissimo sublimbato; interdum areolata, quasi arenacea.
Fig. 510: facies lateralis.
Fig. 511: facies oralis.

Uvig. (Sagr.) dimorpha PARK. & JONES, 1865, North. Atl. & Arct. Oc., Phil. Transact. 155, p. 420, t. 18, f. 18.
Sagr. dimorpha BRADY, 1884, Chall. Rep. 9, p. 582, t. 76, ff. 1—3.

Hab. in sinubus Norvegicis, ut Bukken, Österfjord etc. profund. metr. 260—350 passim (NORMAN), pygmæn, 0.38 mm.

APPENDIX.

U. canariensis D'ORB.

Tab. IX, figg. 489—493.

Lævis, nitida ovata aut clavate fusiformis, obtuse trigona, segmentis plerumque tri-
serialibus plus minusve inflatis.
Fig. 489: minus tumida.
Fig. 490: facies oralis.
Fig. 491: forma tumida.
Fig. 492: facies oralis.

Uv. Canariensis D'ORB., 1839, For. Canar., p. 138, t. 1, ff. 25—27.
> » BRADY, 1884, Chall. Rep. 9, p. 573, t. 74, ff. 1—3.
? **U. nodosa var.** β D'ORB., 1826, Tab. méthod., A. Sc. nat. 7, p. 269, N:o 3.
U. proboscidea SCHWAG., 1866, For. Kar.-Nikob., Novara Reise geol. Th. 2, p. 250, t. 7, f. 96.

Hab. ad Azores insulas profund. metr. 530 sat frequens (SMITT, LJUNGMAN); 1 mm. longa circiter.

Variat minus inflata, triangularis, præterea ut typica. Tab. IX, fig. 493.

Hab. cum typica passim.

U. Auberiana D'ORB.

Tab. IX, figg. 494—495.

Impolita, scabra aut spinosa, segmentis interdum subbiserialibus.
Forma præcedenti similis, sæpe nisi superficie scabra et statura minore diversa.
Fig. 494: facies lateralis.
Fig. 495: facies oralis.

U. **Auberiana** D'ORB., 1839, For. Cuba, p. 106, t. 2, ff. 23—24 (pygmæa).
    GOËS, 1882, Ret. Rhizop. Carib. Seu; Sv. Vet. Akad. Handl. 19, N:o 4, p. 60, t. 4, ff. 71—74
    (pygmæu, spinis sparsis aut deficientibus).
U. **asperula** BRADY, 1884, Chall. Rep., p. 578, t. 75, ff. 6—7 (spinosa).
  »   »   BRADY, PARK., JONES, 1887, Abrohlos Bank, Trans. Zool. Soc. Lond. 12, 7, t. 45, ff. 4—5.
? U. **farinosa** v. HKEN, 1875, For. Cláv. Száb. Sch. (separ.), p. 62, t. 7, f. 6 (spinis destituta).
U. **gracilis** REUSS., 1851, Sept. Thon, Berlin, Zeitschr. d. geol. Ges. 3, p. 77, t. 5, f. 39 (spinosa).

Obs.   Uv. asperulu et Orbiguyana CZJZEK, 1847, For. Wien. Beck., Haid. nat. wiss. Abh. 2, p. 146, t. 13,
    ff. 14—17, U. pygmeæ magis propinquæ videntur.

Hab. cum præcedente ad insulas Azores (SMITT & LJUNGMAN), 0.50—0.70 mm.

CHILOSTOMELLA v. REUSS.

Ch. ovoidea v. RSS.

Tab. IX, figg. 512—516.

Hyalina, ovata, ellipsoidea aut subcylindrica, segmento ultimo, tres-quatuor partes penultimi aut plus obvelante, sutura nunc subhorizontali nunc valde obliquata faciem aboralem versus declinante; poris minutis creberrimis, majoribus sparsis intermixtis (his tamen interdum deficientibus); dispositio segmentorum fere "biloculina".

Figg. 512—513: exemplum sutura fære transversa.
Fig. 514: longitudine secta.
Figg. 515—516: facies orales.

REUSS., 1849, Neue For. Österr., Wien. Ak. DkSchr. 1, p. 380, t. 48, f. 12.
Ch. **Czjzeki** REUSS., ibid., t. 48, f. 13.
Ch. **cylindroides** REUSS., 1851, Septar. Thon Berl. Zeitschr. deut. geol. Gesellsch. 3, p. 80, t. 6, f. 43.
  »   »   **tenuis** BORNEM., 1855, Septar. Th. Hernsdorf., ibid. 7, p. 343, t. 17, f. 1, 2.
1870, v. SCHLICHT, t. 25, ff. 17—22, 25—48.
Ch. **cylindrica** v. HKN, For. Clávul. Szaboi Sch. (separ.) p. 63, t. 7, f. 7.
Ch. **ovoidea** BRADY, 1879, Notes on Ret. Rhizop., Qu. Journ. microsc. Sc. (n. ser.) 19, p. 280, t. 8, ff. 11—12.
  »   »   BRADY, 1884, Chall. Rep., p. 436, t. 55, ff. 12—23.

Hab. in Österfjord Norvegiæ profund. metr. 180—350 (NORMAN). Long. mm. 0.50—0.80.

ALLOMORPHINA v. REUSS.

A. trigona v. REUSS.

Tab. IX, figg. 517—519.

Subtrigona aut trilobata, cordiformis compressa aut subcompressa; dispositio segmentorum "triloculina", ceterum ut in præcedente; apertura rima lateralis, umbilicalis, suturalis inter ultimum et penultimum aut inter ultimum et antepenultimum segmentum instituta. Genus immerito a Chilostomella disjunctum.

Fig. 517: facies lateralis, aboralis, anfractu antepenultimo vix obvelato.
Fig. 518: facies lateralis, oralis.
Fig. 519: facies marginalis.

v. Rss., 1849, Neue For. Österr., Wien. Ak. DkSchr. 1, p. 380, t. 48, f. 14.
Brady, 1879, Notes on Ret. Rhizop., Qu. Journ. micr. Sc. (n. ser.) 19, p. 281, t. 8, ff. 13—14.
  ″ 1884, Chall. Rep., p. 438, t. 55, ff. 24—26.
All. cretacea v. Rss, 1850, Kreidem. Lemberg., Haid. naturw. Abh. 4, p. 42, t. 4, f. 6.
Fés' Mod. 74.

Hab. ad Spetsbergiam in Redbay profund. metr. 20 invenit unicam G. NORDENSKIÖLD. Long. 0.50 mm.

POLYMORPHINA d'Orb.

P. lactea WALK. & JAC.

Late ovalis aut suborbicularis, compressa aut subcompressa, segmentis conspicuis 3—4.

Serpula lactea WALK. & JAC., 1798 (KANMACHER), Adams' Essays, Ed. II, t. 24, f. 4.
Globulina irregularis, æqvalis d'ORB., 1846, Bass. tert. Vienne, p. 227, t. 13, ff. 9 - 12.
    »   inflata Rss, 1851, Septar. Thon. Berlin, Zeitschr. deutsch. geol. Ges. 3, p. 81, t. 6, f. 45.
Guttulina turgida Rss, 1855, Tert. Schicht. nördl. u. mittl. Deutschl., Wien. Ak Sitz. Ber. 18, p. 246, t. 6,
    ff. 66 (a var. gibba d'ORB. haud multum diversa).
Polym. globosa (v. Münst.), Roem., 1838, Nordd. tert. Moerassand, Leonh. u. Bronns Jhb. 1838, p. 386, t. 3, f. 32.
Globul. amygdaloides Rss, 1851, Sept. Thon. Berlin, Zeitschr. deut. geol. Gesellsch. 3, p. 82, t. 6, f. 47.
    »   subalpina GÜMB., 1868, Nordalpin. Eocän., K. Bay. Ak. Wissensch. Abh. 10, 2, p. 646, t. 2, f. 80.
    »   globosa Rss, 1846, Böhm. Kreide. 1, p. 40, t. 13, f. 82; 1861, Kreidetuff Maestrich., Wien. Ak.
    Sitz. Ber. 44, p. 318, t. 3, f. 3.
Polym. Münsteri v. HANTKEN, 1875. Clávul. Szaböi Sch., Szabo., p. 61, t. 7, f. 16.
Glob. Römeri REUSS, 1855, Tert. Sch. u. mittl. Deutschl., Wien. Ak. S.Ber. 18, p. 245, t. 6, f. 63.
Guttul. diluta BORNEM., 1860, For. Magdeburg, Zeitschr. deutsch. geol. Gesellsch. 12, p. 160, t. 6, f. 11.
Polym. ovulum REUSS, 1855, Tert. Sch. nördl. u. mittl. Deutschl., Wien. Ak. Sitz. Ber. 18. p. 250, t. 8, f. 83.
    »   gravis KARR., 1870, Kreidef. Leitzendorf., Jbb. K. K. geol. Reichsanst. Österr. 20, p. 181, t. 11, f. 12.
    »   Münsteri REUSS, 1855, Tert. Sch. nördl. u. mittl. Deutschl., Wien. Ak. S.Ber. 18, p. 249, t. 8, f. 80
    (in P. compressam d'ORB. vergens).
Guttul. Ralbliana GÜMB., 1860, Cassian. u. Raibl. Sch., K. K. geol. Reichsanst. Österr. 19, p. 182, t. 6, f. 31.
    »   mutabilis COSTA, 1854, Pal. Nap. 2, p. 275, t. 18, ff. 1—3 (in var. acutam ROEM. vergens).
    »   deformata REUSS, 1855, Tert. Sch. nördl. u. mittl. Deutschl., Wien. Ak. S.Ber. 18, p. 245, t. 6, f. 64.
?   »   gravida v. compressa TERQ., For. & Ostrac. de l'Islande, Bull. Soc. Zool. Frs. 1886, p. 334, t. 11, f. 21.
Glob. inæqualis Rss, 1849, Neue Foramf. Österreichs, Wien. Ak. DkSchr. 1, p. 377, t. 48, f. 9.
Polym. (Guttulina) compressa d'ORB., 1826, Tabl. method., Ann. Sc. nat. 7, p. 266, t. 12, ff. 1—4, Mod. 62.
Guttul. communis d'ORB., 1846, Bass. tert. Vienne, p. 224, t. 13, ff. 6—8.
Polym. communis BRADY, PARK. & JONES, 1870, Genus Polymorph., Transact. Lin. Soc. 27, p. 224, t. 39, f. 10.
    »   ″   BRADY, 1884, Chall. Rep. 9, p. 568, t. 72, f. 19.
Glob. depauperata Rss, 1867, Steinsaltz. Wieliczka, Wien. Ak. S.Ber. 55, p. 89, t. 3, f. 9 (inter compressam
    & communem).
Polym. communis ROEM., 1838, Leonh. u. Bronns. Jhb. 1838, p. 385, t. 3, f. 29.
Guttul. cretacea ALTH., Umgeb. Lemberg., Haid. naturw. Abh. 3, 2, p. 262, t. 13, f. 14 (in var. problema vergens).
    »   ″   Rss, Kreidemerg. Lemberg., Haid. naturw. Abhandl. 4, 1, p. 44, t. 4, f. 10, t. 13, f. 14 (in
    var. problema d'ORD. vergens).
    »   semiplana Rss, 1851, Sept. v. Berlin., Zeitschr deutsch. geol. Gesellsch. 3, p. 82, t. 6, f. 48.
    »   fissurata, obliqvata STACHE, 1864, Tert. Merg. d. Whaingaron Hafens, Novara Reise geol. Theil 1,
    p. 263, t. 24, ff. 10—11.
Polym. lactea BRADY, PARK. & JONES, 1870, Gen. Polymorph., Transact. Lin. Soc. 27, p. 213, t. 39, f. 1.
    »   ″   BRADY, 1884, Chall. Rep. 9, p. 559, t. 71, f. 11.
    »   amygdaloides BRADY, ibid., p. 560, t. 71, f. 13.
    »   problema v. deltoidea v. HANTK., For. Clàv. Szab. Sch. (Separ.), p. 591, t. 8, f. 3.
    »   lactea WILLIAMS., 1858, Rec. For. gr. Brit., p. 71, t. 6, f. 149.
    »   var. communis WILLIAMS., 1858, ibid., p. 72, t. 6, ff. 153—155.

Obs. Polym. lactea var. oblonga WILL., est varietas distincta.

Hab. in sinubus Bahusiæ et Norvegiæ profund. metr. 60—120 minus frequens.

* forma aulostoma:

Polym. lactea var. tubulosa Park. & Jones, 1865, N. Atl. & Arct., Or., Phil. Transact. 155, p. 362,
t. 13, f. 52.

** rudis:

Polym. ampla Karr., 1870, Kreidef. Leitzerdorf., Jhb. K. K. geol. Reichsanst. Österreich. 20, p. 181,
t. 11, f. 13.

formæ affines:

1. globosa aut late lacrymæformis, aut inflate lenticularis, segmentis apparentibus
3—6, a typica non distincta:

P. gibba d'Orb., Tab. IX, figg. 520—526.

Figg. 520—521: forma gibba d'Orb. fere typica.

Fig. 522: forma "avlostoma" ejusdem.

Figg. 523—524: in "communem" d'Orb. vergens.

Figg. 525—526: magis lenticularis.

Globul. gibba d'Orb., 1826, Tab. méth., Ann. Sc. nat. 7, p. 266, No. 20; 1846, Rass. tert. Vienne,
p. 227, t. 13, ff. 13—14.
Polym. lactea Park. & Jones, 1864, North. Atl. and Arct. Oc., Phil. Transact. 155, p. 359, t. 13,
ff. 45—46.
Glob. amplectens Rss, 1851, Septar. Thon. Berl., Zeitschr. deutsch. geol. Gesellsch. 3, p. 81, t. 6, f. 44,
Polym. gibba var. orbicularis Karr., Mioc. Kostej., Wien. Ak. Sitz. Ber. 58, p. 174, t. 4, f. 8.
Glob. subgibba Gümb., 1868, Nordalp. Eocän, Abb. K. Bayer. Akad. Wissensch. 10, 2, p. 645, t. 2, f. 79.
    » tubulifera, amplectens Bornem., 1860, For. Magdeburg, Zeitschr. deutsch. geol. Gesellsch. 12,
p. 160, t. 6, ff. 10, 12.
    » lacryma Rss, 1846, Böhm. Kreidef. 1, p. 40, t. 12, f. 6; t. 13, f. 83.
    »   » Alth., 1850, Umgeb. Lemberg, Haid. naturwissensch. Abh. 3, 2, p. 263, t. 13, f. 16.

Hab. in sinubus Bahusiæ et Norvegiæ profund. metr. 60—120 passim; forma subcompressa frequentior.

* forma avlostoma:

Glob. tubulosa d'Orb., 1846, Rass. tert. Vienne, p. 228, t. 13, ff. 15—16.
Polym. orbignii Br., Park. & Jones, 1870, Genus Polymorphina, Transact. Lin. Soc. 27, p. 244, t. 13, f. 38.
    » gibba Brady, 1883, Chall. Rep. 9, p. 562, t. 73, f. 16.
    »   » Wright, 1885, Proc. Belfast nat. Field Club 1884—85, Append. t. 26, f. 11.

** rugosa, spinosa aut tuberculata:

Glob. rugosa d'Orb., 1846, Rass. tert. Vienne, p. 229, t. 13, ff. 19—20.
Polym. leprosa Rss, 1867, Steinsaltz abl. Wieliczka, Wien. Ak. S.Ber. 55, p. 89, t. 4, f. 3.
Glob. asperula Gümb., 1868, Nordalp. Eocän, K. Bay. Ak. Wiss. Abh. 10, p. 646, t. 2, f. 81.
Polym. horrida Karr., 1877, Hochqu. Wasserleit, Abh. geol. Reichsanst. Österr. 9, p. 385, t. 16, f. 46.
Glob. spinosa d'Orb., 1846, Rass. tert. Vienne, p. 230, t. 13, ff. 23—24.
Polym.   » Balkw. & Wright, 1885, Rec. Dubl. For., Transact. R. Irish Ac. Sc. 28, p. 347, t. 12, f. 27.
Glob. punctata d'Orb., 1846, Rass. tert. Vienne, p. 229, t. 13, ff. 17—18.
? Polym. foveolata Rss, 1867, Steinsaltzablag. Wieliczka, Wien. Ak. Sitz. Ber. 55, p. 90, f. 2 (Polym.
æquali d'Orb. par).
Polym. asperula Karr., 1877, Hochqu. Wasserleit, Abh. geol. Reichsanst. 9, p. 385, t. 16, f. 47
(formæ typicæ proximans).
    » hirsuta Brady, Park., Jones, 1870, Gen. Polymorph., Transact. Lin. Soc. 27, p. 243, t. 42, f. 37.
Glob. tuberculata d'Orb., 1846, Rass. tert. Vienne, p. 230, t. 13, ff. 21—22.
Polym.   » Brady, Park., Jones, 1870, Gen. Polymorph., Transact. Lin. Soc. 27, p. 242, t. 41, f.36.

2.   magis producta, ovalis, fusiformis, subcylindrica aut pyriformis, teres aut subcompressa, segmentis apparentibus circiter 3—5.
A Pyrulina gutta D'ORB. vix distincta.

P. acuta ROEMER. Tab. IX, figg. 527—528, Tab. X, figg. 537—538.

Figg. 527—528: P. sororia v. REUSS, e mari Spetsbergico.
Figg. 537—538: producte ovalis subcompressa, e mari Newfoundlandico.

Globulina acuta ROEMER, 1838, Tert. Meeress, LEONH. n. BRONNS Jhb. 1838, t. 3, f. 36.
Polym. clavata ROEMER, ibid., t. 3, f. 38.
Glob. acuta v. REUSS, 1855, Tert. Sch. nördl. u. mittl. Deutschl., Wien. Ak. Sitz. Ber. 18, p. 245, t. 6, f. 62.
Polym. longicollis KARR., 1870, Kreidef. Leitzendorf., Jhb. K. geol. Reichsanst. Österr. 20, p. 181,
    t. 11, f. 11.
Glob. guttula Rss, 1851, Sept. Thon. Berl., Zeitschr. deutsch. geol. Gesellsch. 3, p. 82, t. 6, f. 46.
    »   minuta Rss, 1849, Neue For. Österr., Wien. Ak. DkSchr. 1, p. 377, t. 48, f. 8.
Polym. oblonga, minuta ROEM., 1838, Nordd. tert. Meeress., LEONH. u. BRONNS Jhb. 1838, p. 386,
    t. 3, ff. 34—35.
Guttul. deplanata Rss, 1855, Tert. Sch. nördl. u. mittl. Deutschl., Wien. Ak. Sitz. Ber. 18, p. 246, t. 6, f. 67.
Globul. lacryma Rss, 1850, Kreidem. Lemberg, Haid. nat. Abh. 4, p. 43, t. 4, f. 9.
Polym. amygdaloides Rss, 1855, Tert. Schicht. nördl. u. mittl. Deutschl., Wien. Ak. Sitz. Ber. 18,
    p. 250, t. 8, f. 84 (in P. compressam verg.)
    »   sororia Rss, 1863, Septar. Thon. Offenb., Wien. Ak. Sitz. Ber. 48, p. 57, t. 7, ff. 72—74.
    »   »   Rss, For. Crag. d'Anvers, Bull. Ac. Belg. (2) 15, p. 151, t. 2, ff. 25—29.
Globul. minima BORNEM., 1855, Sept. Thon. Hermsdorf, Zeitschr. deutsch. geol. Gesellsch. 7, p. 344,
    t. 17, f. 3.
Guttul. austriaca D'ORB., 1846, Bass. tert. Vienne, p. 223, t. 12, ff. 23—25 (ad P. problema a
    BRADY relata).
? Glob. prisca Rss, 1862, Nordd. Hils u. Gault, Wien, Ak. Sitz. Ber. 46, p. 79, t. 9, f. 8.
Polym. liassica STRICKL., 1846, Qvart. Journ. geol. Soc. 2, p. 30, f. a.
Polym. similis Rss, 1855, Tert. Sch. nördl. u. mittl. Deutschl., Wien. Ak. Sitz. Ber. 18, p. 249, t. 7, f. 79.
? Polym. amœna KARR., 1877, Hochqu. Wasserl., Abb. K. geol. Reichsanst. Österr. 9, p. 385, t. 16,
    f. 45 (in Pol. lanceolatam Rss transiens).
Polym. lactea, sororia, sororia v. cuspidata BRADY, 1884, Chall. Rep. 9, p. 559, 562, 563, t. 71,
    ff. 14—19, t. 72, f. 4.
a Polym. fusiformis ROEM. et Globul. porrecta REUSS, Westphal. Kreid., Wien. Ak. Sitz. Ber. 40,
    p. 230, t. 12, f. 4, non aut. distincta.

Hab. ad oras Newfoundlandiæ, Spetsbergiæ profund. metr. 80—180 passim.

* avlostoma:

Globul. horrida Rss, 1846, Böhm. Kreid. 2, p. 110, t. 43, f. 14; Kreidem. Lemberg, Haid. nat. Abh. 4,
    p. 43, t. 4, f. 8.
Avlostomella pediculus ALTH., 1849, Umgeb. Lemberg, Haid. nat. Abh. 3, p. 264, t. 13, f. 17.
Polym. sororia BRADY, 1884, Chall. Rep. 9, p. 562, t. 73, f. 15.

3.   subcompressa aut teres, segmentis magis inflatis, extra apparentibus 5—7; a
typica non distingvenda.

P. problema D'ORB.

Guttulina problema D'ORB., 1826, Tab. méth., Ann. Sc. nat. 7, p. 266, No. 14, Mod. 61; 1846,
    Bass. tert. Vienne, p. 224, t. 12, ff. 26—28.
Polymorph. crassatina (v. MÜNST.), spicœformis ROEM., 1838, Nordd. tert. Meeressand, LEONH. u.
    BRONNS Jhb. 1838, p. 385, t. 3, ff. 30—31.

Polym. problema BRADY, PARK., JONES, 1870, Gen. Polymorph., Trans. Lin. Soc. 27, p. 225, t. 39, f. 11; Crag. lorrmf. Pal. Soc. 19, t. 1, f. 64; BRADY, 1884, Chall. Rep., p. 568, t. 72, f. 20, t. 73, f. 1.
Polym. problema, uvula Egg., 1857, Mioc. Ortenburg, LEONH. n. BRONNS Jhb. 1857, p. 285, t. 10, ff. 23—29.
Gutt. rotundata, problema Rss, 1864, Oberoligocän, Wien. Ak. Sitz. Ber. 50, p. 469, t. 3, f. 4, p. 470 t. 5, f. 5
» **pusilla** STACHE, 1865, Whaingaroa Haf., Novara Reise geol. Th. 1, p. 264, t. 24. f. 12.
Polym. acuminata v. HANTK., 1875, For. Clav. Szaböi Sch., p. 60, t. 8, f. 4.
Glob. discreta Rss, 1849, Neue For. Österr., Wien. Ak. DkSchr. 1, p. 378, t. 48, f. 10.
? **Gutt. vitrea** D'ORB., 1839, For. Cuba, p. 133, t. 3, ff. 1—3 (sede incerta).

Hab. e. præcedente rara.

* rudis:

Glob. caribæa D'ORB., 1839, For. Cuba, p. 135, t. 2, ff. 7—8.

P. rotundata BORNEM.

Tab. IX, figg. 529—534.

Ovalis aut pyriformis, teres aut subcompressa, segmentis apparentibus 4—8 irregulariter dispositis.

Variat: subcylindrica, segmento adultarum ultimo margine spinis armato, Nodosariæ armatæ REUSS similis.

Figg. 529—530: e Korsfjord, Norvegiæ.

Fig. 531: altera ex eodem loco paullum compressa.

Fig. 532: facies marginalis ejusdem.

Figg. 533—534: var. armata, e fretu Koster insularum.

Guttul. rotundata BORNEMANN, 1855, Septar. Thon. Hermsdorf, Zeitschr. deutsch. geol. Gesellsch. 7, p. 346, t. 18, f. 3.
Guttul. fracta, Globul. dimorpha, incurva, globosa, obtusa BORNEM., ibid., p. 344—346, t. 17, ff. 4—6, t. 18, ff. 1—2.
Polym. rotundata Rss, 1870, v. SCHLICHT, For. Pietzpuhl., t. 26, ff. 13—15, t. 28, ff. 1—5, t. 30, ff. 33—40.
» **turgida** Rss, v. SCHLICHT, ibid., t. 28, ff. 6—10, t. 29, ff. 1—5.
» **tenera** KARR., 1868, Mioc. Kostej; Wien. Ak. S.Ber. 58, p. 174, t. 4, f. 9.
» **rotundata** BR., PARK. & JONES, 1870, Genus Polymorph., Transact. Lin. Soc. 27, p. 234, t. 40, f. 19, a—c.
» **gigas** KARR., 1877, Hochquell. Wasserleit., Abh. geol. Reichsanst. Österr. 9, p. 384, t. 16, f. 44.
» **rotundata** BRADY, 1884, Chall. Rep. 9, p. 570, t. 73, ff. 5—8.

Ab hac non sat distincta:

Polym. oblonga D'ORB., 1846, Bass. tert. Vienne, p. 332, t. 12, ff. 29—31; BRADY, 1884. Chall. Rep., p. 569, t. 73, ff. 2—4.
Polym. uvæformis Rss, 1855, Kreidemerg. Mecklenb., Zeitschr. deutsch. geol. Gesellsch. 7, p. 289, t. 11, f. 5
? **Polym. Soldanii** D'ORB., 1826, Tabl. méth., Ann. Sc. Nat. 7, p. 265, No. 12.
? **Polym. subdepressa** Rss, 1855, Tert. Sch. nördl. u. mittl. Deutschl., Wien. Akad. S.Ber. 18, p. 249, t. 8, f. 81.
Polym. leopolitanæ Rss, 1850, Kreidemerg. Lemberg, Haid. naturw. Abh. 4, p. 44, t. 4, f. 11.
? **Polym. rudis** Rss, 1861, Kreidetuff. Maastrich., Wien. Ak. S.Ber. 44, p. 319, t. 3, ff. 5—8.

Hab. in sinubus Bahusiæ et Norvegiæ profund. metr. 90—360 passim.

P. angusta Egger.

Tab. X, figg. 535—536.

Subcylindrica aut elongate ovata, teres aut subteres, segmentis apparentibus 4—6. A Polym. lactea var. acuta Roem., cujus forma pygmæa, non sat distincta.

Glob. angusta Egger, 1857, Mioc. Ortenburg, Leonh. & Bronns Jhb. 1857, p. 290, t. 13, ff. 13—15.
Polym. fusiformis (ex parte) Brad., Park., Jones, 1870, Genus Polymorph., Trans. Lin. Soc. 27, p. 219.
      angusta Brady, 1884, Chall. Rep. 9, p. 563, t. 72, ff. 1—3.

Hab. mare Spetsbergicum profund. metr. 30 rara; pygmæa, long. mm. 0,60.

P. compressa d'Orb.

Tab. X, figg. 539—553.

Plus minusve compressa, late aut elongate ovalis, segmentis irregulariter biserialibus apparentibus 5—8, plerumque non inflatis, suturis valde obliquis aut magis transversis. Valde varians segmentis interdum inflatis, a Polymorph. ancipite Phil. et complanata d'Orb. difficulter distincta; in Polym. lacteam sæpe mergens.

Figg. 539—544: formæ variæ e mari Norvegico.

Figg. 545—546: e sinu Codano.

Figg. 547—549: e mari Norvegico.

Figg. 550—551: e mari Spetsbergico.

Figg. 552—553: e mari Atlantico extra Azores insulas.

Polym. compressa d'Orb., 1846, Bass. tert. Vienne, p. 233, t. 12, ff. 32—34 (valde elongata).
? **Polym. tuberosa** d'Orb., 1826, Tab. méth., Ann. Sc. Nat. 7, p. 265, No. 6.
Polym. lactea Will., 1858, Brit. Rec. For., p. 70, t. 6, ff. 145—146.
? **Polym. racemosa** Terqu., 1886, For. etc. de l'Islande, Bull. Soc. Zool. Fr. 1886, p. 335, t. 11, f. 22.
Polym. pygmæa Schwag., 1863, Jurass. Sch., Würtemberg natwiss. Jahrb. 21, p. 138, t. 7, f. 8.
    » **lactea** var. **compressa** Park. & Jones, 1865, North Atl. & Arct. Oc., Phil. Transact. 155, p. 361, t. 13, ff. 47, 49, 51.
    » **compressa** Brady, Park., Jones, 1870, Genus Polymorph., Trans. Lin. Soc. 27, p. 227, t. 40, f. 12 a—f
    »   » Brady, 1884, Chall. Rep. 9, p. 565, t. 72, ff. 9—11.
    » **lingva** Roem., 1838, Nordd. tert. Meeress., Leonh. & Bronn Jhb., p. 385, t. 3, f. 25.
Globul. discreta, Guttul. robusta Rss, 1864, Oberoligocän, Wien. Ak. Sitz. Ber. 50, pp. 468, 470, t. 3, ff. 3, 5—7.
Polym. sacculus, incavata Stache, 1865, Novara Reise geol. Th. 1, 2, p. 259, 260, t. 24, ff. 6—7.
? **Guttulina elliptica** Reuss, 1845, Böhm. Kreidef. 2, p. 110, t. 24, f. 55.
Polym. Schwageri Karr., 1877, Hochquell. Wasserleit, Abh. geol. Reichsanst. Österr. 9, p. 384, t. 16, f. 43.
Polym. Zeuschneri Rss, 1867, Steinsaltzablag. Wieliczka, Wien. Ak. Sitz. Ber. 55, p. 90, t. 4, f. 1.
    » **compressa** Brady, Park. & Jones, Crag. For. 19, t. 1, ff. 54, 55, 77 (segmentis transversis).
    » **subrombica** Rss, 1861, Grünsand Newjersey, Wien. Ak. Sitz. Ber. 44, p. 339, t. 7, f. 3 (latior.)
    » **insignis** Phil., Reuss. 1855, Tert. Sch. nördl. u. mittl. Deutschl., Wien. Ak. Sitz. Ber. 18, p. 248, t. 7, ff. 74—76.
    » **dispar** Stache, 1865, Nov. Reise geol. Th. 1, 2, p. 261, t. 24, f. 8 (lata).
    » **Humboldti** Bornem., 1855, Septar. Thon. Hermsd., Zeitschr. deutsch. geol. Gesellsch. 7, p. 347, t. 18, ff. 7—8 (lata).

Ab his non sat distincta:

Polym. Burdigalensis d'Orb., 1826, Tab. méth., Ann. Sc. nat. 7, p. 265, No. 2, Mod. 29.
>   >   Brady, Park. & Jones, 1870, Transact. Lin. Soc. 27, p. 224, t. 39, f. 9.

* avlostoma:

Polym. lactea v. **fistulosa** Will., 1858, Rec. For. Gr. Brit., p. 72, t. 6, f. 150.
    Orbignii (ex parte) Brady, Park. & Jones, 1870, Gen. Polymorph., Trans. Lin. Soc. 27, p. 244
         t. 42, f. 38 d.

Hab. in sinubus Bahusiæ et Norvegiæ, nec non Spetsbergiæ profund. metr. 50—180 passim.

P. Thouini d'Orb.

Tab. X, figg. 557—558.

Fusiformis, compressa aut subcompressa, segmentis apparentibus 5—10, irregulariter alternantibus, suturis plus minusve oblique dispositis.

Polym. Thouini d'Orb., 1826, Tab. méth. Ann. Sc. nat. 7, p. 265, No. 8, Mod. 23 (anguste subcylindrica
         segmentis paucis).
>   Brady, Park. & Jones, 1870, Genus Polymorphina, Trans. Lin. Soc. 27, p. 232, t. 40, f. 17
>   >   Brady, 1884, Chall. Rep. 9, p. 567, t. 72, f. 18.

Hab. ad oras Nowaja Semjla profund. metr. 25 rara (Théel et Stuxberg). Long. mm. 2,45.

APPENDIX.

P. irregularis d'Orb.

Tab. X, figg. 554—556.

Ovalis aut subfusiformis, segmentis irregulariter dispositis, inflatis, externe apparentibus 4—7—8, plus minusve striatis; a P. lactea v. problemate nisi striis non diversa. Fig. 554: summa deorsum versa.

Polymorph. irregularis d'Orb., 1839, For. Cuba, p. 137, t. 2, ff. 12—13.
P. regina Brady, Park., Jones, 1870, Genus Polymorph.; Transact. Lin. Soc. 27, p. 241, t. 41, f. 32; Brady,
         1884, Chall. Rep. 9, p. 571, t. 73, ff. 11—13.
P. semicostata Marss., 1877, Rügen. Schreibkr., Greifswald. Mittheil. nat. Verein d. Vorpommern, 1877—78,
         p. 150, t. 2, f. 19.
? **P. australis** (d'Orb.) Brady, Park., Jones, 1870, Genus Polymorph., Trans. Lin. Soc. 27, p. 239, t. 41, f. 27.

Hab. ad Azores ins. profund. metr. 100 rara (Smitt & Ljungman).

NODOSARINA (Lamck.) Park. & Jones.

* Cristellaria d'Orb.

C. rotulata Lmk.

Tab. X, figg. 559—578.

Helicostegica, nautiloidea, lenticularis, marginata, carinata aut carinato-alata, nunc umbonata, nunc umbonibus distinctis destituta. Nomen "cultrata" est ineptum, quia forma fere quæque suam formam alatam sive cultratam habet. Forma compressa plerumque "cultrata" sæpeque magis polystegica, in Crist. reniformem d'Orb. transiens. Multitudine nominum gravata.

Figg. 559—562: exempla minuta e mari Azorico; in articulatam v. Reuss vergentia.

Figg. 563—564: exemplum magis polystegicum e freto Koster insularum profund. metr. 150.

Figg. 565—566: anguste carinata (inornata, austriaca D'Orb.) e mari Azorico, profund. metr. 230—500.

Figg. 567—568: valida, umbone applanato (Beyrichi Bornemann, declivis Rss.) e mari Spetsbergico profund. metr. 400.

Figg. 569—570: e Skagerrack profund. metr. 270, carinato-"cultrata".

Figg. 571—572: e mari Azorico, profund. metr. 500, cultrata (Rob. canariensis D'Orb.)

Figg. 573—574: elatior; cultrata, e mari Caraibico profund. metr. 540.

Figg. 575—576: magis compressa, cultrata, ad C. Josephinæ Mariæ D'Orb. tendens, e mari Norvegico profund. metr. 350.

Figg. 577—578: compressa, cultrata, magis polystegica; umbone reducto, e mari Azorico profund. metr. 300—500.

Lenticulites rotulata Lmk., (1804) 1827, Tab. Encycl. Méth. Vers., t. 466, f. 5.
Nodos. calcar. Goès, 1882, Ret. Rhizop. Carib. Sea; Sv. Vet. Akad. Handl. 19, 4, t. 3, ff. 55, 57—61 (cultrata et carinata).
Crist. rotulata Brady, 1884, Chall. Rep. 9, p. 547, t. 69, f. 13.

Hab. in fretis Koster insularum non frequens, in sinu Skagerack, in sinubus Norvegicis, in mari Spetsbergico, mari Azorico etc. Diam. usque ad mm. 4,25.

formæ affines:

1. lenticularis lævis, margine dentato-alato aut spinis paucis plus minusve elongatis instructo.

C. calcar Lin.

Nautilus calcar Lin. (ex parte), 1758, Syst. Nat. 10, p. 709.
Naut. calcar var. α, ϑ, χ, μ Ficht. & Moll, 1803, Test. microscop. pp. 71, 76, 79, t. 11, ff. a, b, c, t. 12, ff. i, k, t. 13, ff. c, d, h, i.
Robulina radiata, pulchella, lævigata, aculeata D'Orb., 1826, Tab. méth., Ann. Sc. nat. 7, p. 288, 289, No. 7, 8, 9, 14.
Rob. calcar D'Orb., 1846, Bass. tert. Vienne, p. 99, t. 4, ff. 18—20.
Crist. calcar Brady, 1884, For. Chall. Rep. 9, p. 551, t. 70, ff. 9—15.
»　　» Brady, Park., Jones, 1887, Abrohlos Bank, Trans. Zool. Soc. Lond. 12, 7, t. 44, f. 14.
Nodosarina calcar Goès, 1882, Ret. Rhizop. Carib. Sea, Sv. Vet. Akad. Handl. 19, 4, t. 3, ff. 54—56.

Hab. ad Koster insulas profund. metr. 100, rara atque minuta.

APPENDIX.

2. margine rotundato vel obtuso interdum obtuso-carinato, segmentis paullum inflatis, umbone sæpe vitreo; forma male distincta.

C. articulata BRADY, Tab. X, figg. 579—582, ? 583—584.

Figg. 579—580: e mari Azorico profund. metr. 700.

Figg. 581—582: ex eodem mari profund. metr. 180; in typicam vergens.

Figg. 583—584: e mari boreali Atlantico profund. metr. 1750; minus typica (an Cr. gibbosa D'ORB., Cr. obliqua v. Hagen.).

Crist. articulata BRADY, 1884, For. Chall. Rep. 9, p. 547, t. 69, ff. 10—12.
? Rob. articulata Rss, 1863, Sept. Thon. v. Offenbach, Wien. Ak. S.Ber. 48, p. 53, t. 5, f. 62; v.
SCHLICHT, 1870, tab. 17, ff. 5—12 (carinata).

Hab. ad Azores profund. metr. 180—700 (SMITT & LJUNGM.)

/a 3.   valde inflata sæpe grinata, segmentis paucis, forsan megasphærica formæ typicæ.

C. crassa D'ORB., Tab. X, figg. 585—586.

Crist. crassa D'ORB., 1846, Bass. tert. Vienne, p. 90, t. 4, ff. 1—3.
Rob. deformis Rss, 1851, Sept. Thon. Berlin, Zeitschr. dentsch. geol. Gesellsch. 3, p. 70, t. 4, f. 30.
Crist. crassa BRADY, 1884, For. Chall. Rep. 9, p. 549, t. 70, f. 1.

Hab. ad Azores profund. metr. 100—400 (SMITT & LJUNGMAN).

4.   magis oblonga, interdum compressa et marginato-alata.

C. gibba D'ORB., Tab. X, figg. 587—592.

Figg. 587—588: fere typica (= C. nuda, spectabilis v. Rss).

Figg. 589—590: elatior; Rob. latæ Rss, Crist. reniformi D'ORB. C. Hantkeni RHZEHAK proxima.

Figg. 591—592: valde alata; C. Josephinæ-Mariæ D'ORB.,]C. osnabrügensi v. /nitida..ORB REUSS propinqua.

Crist. gibba D'ORB., 1839, For. Cuba, p. 63, t. 7, ff. 20—21.
Crist. galeata v. REUSS, 1851, Septar. Thon. Berlin (ex parte), Zeitschr. deutsch. geol. Gesellsch. 3,
p. 66, t. 4, f. 20.

Proximæ sunt non sat distinctæ:

Rob. princeps, Kubingii v. HANTK., 1875, For. Cláv. Szab. Sch., p. 56, t. 6, ff. 7—8.
Crist. subcostata Rss, 1855, Tert. Schicht. nördl. u. mittl. Deutschl., Wien. Ak. Sitz. Ber. 18, p. 237,
t. 3, f. 43 (limbata).
? Nautilina putsolana COSTA, 1854, Pal. Nap. 2, t. 27, f. 28.
Crist. Hantkeni RHZEHAK, 1885, Verh. Brünn. Nat. Forsch. Vereins 24, p. 100, t. 1, f. 8.
    » Russeggeri, rostrata REUSS et aliæ.
    » gibba BRADY, 1884, Chall. Rep. 9, p. 546, t. 69, ff. 8—9.

Hab. mare Azoricum profund. metr. 1,400 (SMITT & LJUNGM.); mare Spetsbergicum profund. metr. 450
(in C. rotulatam vergens).

C. variabilis v. Rss.

Tab. X, figg. 593—595.

Helico-stichostegica, plerumque compressa, carinata, apertura subtubulari; minuta.

Crist. **variabilis** v. Reuss, 1849, Neue For. Österr., Wien. Ak. DkSchr. 1, p. 369, t. 46, ff. 15—16 (non sat typica).
>   **perogrina** Schwag., 1866, Novara Exped. geol. Th. 2, p. 245, t. 7, f. 89.
>   **variabilis** Brady, 1884, For. Chall. Rep. 9, p. 541, t. 68, ff. 11—16.
>   >   Brady, Park., Jones, 1887, Abrohlos Bank, Trans. Zool. Soc. Lond. 12, 7, t. 44, f. 12.

Hab. mare Azoricum metr. 230, Smitt & Ljungm. Long. 0,70 mm.

C. crepidula Ficht. & Moll.

Tab. XI, figg. 596—613.

Segmentis juvenilibus helicostegicis aut subhelicostegicis, reliquis oblique stegostegicis; nunc obtuse marginata, nunc carinato-alata.

Figg. 596—611: formæ variæ e mari Azorico.

Figg. 612—613: e mari Atlantico boreali.

Naut. **crepidula** Ficht. & Moll, 1803, Test. microsc., p. 107, t. 19, ff. g—i.
Crist. **Schlönbachi, crepidula** Brady, 1884, For. Chall. Rep. 9, pp. 539, 542, t. 67, ff. 7, 17, 19, 20,
     t. 68, ff. 1—2.
>   **crepidula** Wright, 1885, Chalk. Foram. Keady Hill, Proc. Belfast Nat. Field Club 1884—85, Append. t. 27, f. 4.
>   Balkw., Millett, 1884, For. Galway, Journ. micr. nat. Sc. 3, t. 4, f. 8.
>   >   Brady, Park., Jones, 1887, Abrohlos Bank, Trans. Zool. Soc. Lond. 12, 7, t. 44, ff. 8, 9.
>   >   Fornasini, 1890, Lagen. pliocen. Catanzarese; Mem. R. Accad. Sc. Instit. Bologna (4) 10,
     t. 1, ff. 31—33.
Synonymiam ceteram vide Goës Rhizop. Carib. Sea, Sv. Vet. Akad. Handl. 19, 4, p. 43. Formæ aliquæ inter
     Nod. legumen var. glabram et lituum, p. 35—36 potius ad Nod. crepidulam sunt adscribendæ.
Valde varians, inter varietates diversas nulli fines delineari possunt.

Hab. mare Atlanticum boreale profund. metr. 1,750 rara; Spetsbergense metr. 180 rara; Azoricum metr. 530
     sat frequens (Smitt, Ljungman).

formæ affines:

1.   leguminoides, tota fere stichostegica, interdum valde angustata.
A Marg. Webbiana d'Orb. (ex parte) vix distincta.

C. complanata v. Reuss, Tab. XI, figg. 616—622.

Figg. 616—620: formæ "leguminis".

Figg. 621—622: in C. cymboidem d'Orb., C. nummuliticam Gümb., C. insolitam Schwag. vergens; omnes e mari Azorico profund. metr. 530.

C. **complanata** v. Reuss, 1846, Böhm. Kreideform. 1, p. 33, t. 13, f. 54.
? Crist. **compressa** d'Orb., 1846, Bass. tert. Vienne, p. 86, t. 3, ff. 32—33.
Crist. **perobliqua, incurvata** v. Reuss et permulta nomina alia.
Vagin. legumen Brady, 1884 (partim), Chall. Rep., t. 66, f. 13.

Obs. Crist. compressa Brady, Chall. Rep. 9, t. 114, ff. 15—16 non exacte C. compressam
     d'Orb. exhibet, sed potius C. reniformem d'Orb.

Hab. mare Atlanticum boreale profund. metr. 1,400 rara; mare Azoricum, sat frequens.

2. carinata aut carinato-alata, stadio juvenili spinis duabus prædito marginalibus.

C. spinigera BRADY, Tab. XI, ff. 614—615.
Vagin. spinigera BRADY, 1884, Chall. Rep. 9, p. 631, t. 67, ff. 13—14.

Hab. mare Atlant. boreale profund. metr. 1150 rara (Jos. LINDAHL).

3. elongata, nunc recta nunc curvata, plus minusve complanata, interdum carinato-alata. A typica non sat distincta, aliquando ut varietas Crist. rotulata aptius collocanda videtur; sæpe megasphærica.

C. subarcuatula MONTAG. Tab. XI, figg. 630—637.

Figg. 630—631: exemplum e sinu Skagerack.
Figg. 632—633: e Söderfjord Norvegiæ, profund. metr. 450.
Figg. 634—635: e mari Atlantico boreali profund. metr. 530.
Figg. 636—637: e mari extra Azores insulas, profund. metr. 500 (SMITT & LJUNGMAN).

Nautilus subarcuatulus MONTAG., 1808, Test. Brit. Supplem., p. 80, t. 19, f. 1 (limbata, minuta).
Crist. subarcuatula var. WILLIAMS, 1858, Rec. For. Gr. Brit., t. 2, f. 62.
  »   calcar var. marginulinoides PARK. & JONES, 1857, For. Coast Norway, An. Mag. Nat. Hist.
    (2) 19, p. 269, t. 10, f. 1.
Margin. lituus PARK. & JONES, 1865, North. Atl. & Arct. Oc.; Phil. Transact. 155, p. 343, t. 13, f. 14.
Crist. obtusata var. subalata BRADY, 1884, Chall. Rep. 9, p. 536, t. 66, ff. 24—25.
  »   elongata BRADY, 1887, Synops. Brit. Rec. For., Journ. Microsc. Soc. 1887, p. 911 (nomen a D'ORB.
    pro aliis formis usitatum).

Obs. Crist. obtusata v. REUSS, 1870; v. SCULICHT, Septar. Thon. Pietzpuhl, t. 11, ff. 16—18 ad seriem Cristellariarum tumidarum pertinet, a Crist. articulata v. REUSS et C. Hauerina D'ORB. descendentium.

Hab. Skagerack profund. metr. 100; sinus Norvegicos profund. metr. 90—450; mare Atlant. boreale profund. metr. 530 passim; procera.

APPENDIX.

4. elongata, plus minusve angustata, tumidiuscula, interdum compressa, nunc carinata, nunc margine obtuso aut rotundato, suturis sæpe sublimbatis; aliquando obsolete striolata. Stadium juvenile complete helichostegicum, quod a Crist. rotulata LMK. vix discerni possit.

C. Saulcyi D'ORB., Tab. XI, figg. 623—629.

Figg. 623—627: formæ plus minusve compressæ.
Figg. 628—629: stadium juvenile, magis dilatatum.

Crist. Saulcyi D'ORB., 1839, For. Isles Canaries, p. 126, t. 9, ff. 7—9.

Huic maxime affinis:
Marginulina ensis v. RSS, 1846, Böhm. Kreideform. 1, p. 29, t. 12, ff. 13, t. 13, ff. 26—27 et aliæ.
Hab. mare Azoricum profund. metr. 530 (SMITT & LJUNGMAN).

4. brevis, inflata, plerumque triquetra, obliquo helicostegica, carinato-marginata, spira brevissima. Longitudo et amplitudo spiræ valde varians; a Crist. italica Defr. difficillime limitanda.

C. navicula D'Orb., Tab. XI, figg. 638—640.

Figg. 638: facies lateralis.
Figg. 639: facies marginalis sive dorsalis.
Figg. 640: facies spiralis sive ventralis.

Crist. navicula D'Orb., 1839, Craie bl. Paris, Mém. Soc. géol. Fr. 4, p. 27, t. 2, ff. 19—20.
　　　　　» triangularis d'Orb., ibid., p. 27, t. 2, ff. 21—22.
? Naut. acutauricularis F. & M., 1803, Test. micr., p. 102, t. 18, ff. g—i (limbata).
Crist. arcuata D'Orb., 1846, Bass. tert. Vienne, p. 87, t. 3, ff. 34—36.
　　» italica D'Orb. (partim), Mod. 19.
　　» acutauricularis, latifrons Brady, 1884, For. Chall. Rep. 9, pp. 543, 544, t. 114, f. 17, t. 68, f. 19, t. 113, f. 11.

Hab. mare Azoricum metr. 530 (Smitt & Ljungman).

5. lineolata aut subcostata, sæpe seminuda, sublimbata, marginata, compressa aut tumidiuscula, Marginul. rugoso-striatæ Gümb. et Marg. tonsillari Gümb. propinqua.

C. elegantissima Costa, Tab. XI, figg. 641—642.

Figg. 641, 641 b: facies lateralis.
Fig. 642: facies spiralis-marginalis.
Fig. 642 a: facies oralis.

Robulina elegantissima Costa, 1854, Pal. Napoli 2, p. 198, t. 19, f. 4 (minus compressa, sublimbata).
? Crist. bicostata Deecke, 1884, Abh. geol. Spec. Kart. Elsass-Lothringen 4, p. 49, t. 2, f. 13 (valde compressa, limbata).

Hab. mare Azoricum profund. metr. 530. Long. m. 1,68.

6. statura præcedentis, stadio maturo tereti aut subtrigono, costata aut striata, interdum carinata, apertura sæpe protrusa. Margin. costatæ Batsch quam maxime propinqua, forsan microsphærica ejusdem.

C. Bradyi Goés, Tab. XI, figg. 643—645 a.

Figg. 643, 645: facies lateralis.
Fig. 644: facies marginalis spiralis.
Fig. 645 a: facies oralis.

Marginulina costata Brady, 1884, Chall. Rep. 9, t. 65, f. 11.

Hab. mare Caraibicum profund. metr. 830, rara. Long. mm. 1—1,25.

** Vaginulina.

V. lævigata Roem.

Tab. XI, figg. 646—655.

Tota stichostegica, segmentis sæpe minus compressis, subteretibus aut stadio juvenili solum compresso, stadio maturo teretiusculo. Formæ tenues fere teretes a Nod. communi d'Orb. sive inornata d'Orb. vix limitandæ. Conf. Goës, Ret. Rhizop. Carib. Sea, Sv. Vet. Akad. Handl. 19, 4, t. 2, ff. 22, 25 secundum Brady ad Nod. communem d'Orb. referendæ. Nodos. legumen v. Rss. Böhm. Kreidef. ad hanc formam subcompressam pertinet.

Figg. 646—647: exempla in Marg. glabram vergentia, e mari Caraibico, stadio juvenili (a) compressiusculo, stadio maturo (b) subtereti.

Fig. 648: pygmæa, in C. crepidulam vergens.

Figg. 649—650: formæ pygmææ, omnes e mari Caraibico profund. metr. 500.

Figg. 651—652: e mari Spetsbergico profund. metr. 1,600.

Fig. 653: e fretis Koster insularum profund. metr. 100.

Fig. 654: e mari extra Azores insulas profund. metr. 530.

Fig. 655: in sequentem vergens, ex eodem mari profund. metr. 180.

Vaginulina lævigata Roem., 1838, Norddeutsch. tert. Meeress; Leon. & Bronns Jhb. 1838, p. 383, t. 3, f. 11.
Nautilus legumen testa recta compressa articulata hinc marginata, siphone laterali. Lin., 1758, Syst. Nat. Ed. X, 1, p. 711 (etiam Vag. elegans d'Orb.).
Dental. legumen Williams, 1858, Rec. For. Gr. Brit., p. 22, t. 2, f. 45.
Vaginulina legumen Brady, 1884, Chall. Rep. 9, p. 530, t. 66, ff. 14—15.

Hab. ad Koster insulas profund. metr. 100, rara, pygmæa; mare Spetsbergicum profund. metr. 1,600; ad Azores insulas profund. metr. 500 (Smitt & Ljungman).

formæ affines:

1. tota teres aut subcompressa, angustata aut subinflata, suturis obliquis aut subtransversis. A typica nec a Marg. regulari d'Orb. nec a Nodos. Ræmeri Neugeboren sat distincta:

V. glabra d'Orb. Tab. XI, figg. 656—661.

Fig. 656: e mari Atlantico extra Azores insulas; profund metr. 80—150; a: facies oralis.

Fig. 657: e sinu Bergensium Norvegicorum profund. metr. 100; a: facies oralis.

Fig. 658: e mari Bahusiæ extra Hållö insulam profund. metr. 45. Nod. Ræmeri proxima.

Fig. 659: fere typica (= Margin. Bertheloti d'Orb.; Marg. subbullata v. Hken; Marg. bullata Rss etc.) e mari Azorico profund. metr. 220; a: facies oralis.

Figg. 660—661: angustior, eodem ex loco.

Marginulina glabra D'ORB., 1826, Tabl. Méth., Ann. Sc. Nat. 7, p. 259, No 6; Mod. 55.
Marg. Berthelotiana D'ORB., 1839, For. Iles Canaries, p. 125, t. 1, ff. 12—13.
   » **glabra** BRADY, 1884, Challeng. Rep. 9, p. 527, t. 66, ff. 5—6 (valde incrassata).
      » FORNASINI, 1890, Lagen. plioc. Catanzarese; Mem. R. Accad. Sc. Istit. Bologna (4) 10,
                      t. 1, ff. 26—30.
? Crist. irregularis v. HKEN, 1876, For. Clav. Szab. Schicht. (sep.), p. 50, t. 14, f. 3 (compressa).

Hab. mare Atlant. boreale profund. metr. 1,200; sinus Norvegiæ profund. metr. 100; mare Bahusiæ;
    mare Azoricum profund. metr. 150—500 (SMITT & LJUNGMAN).

2. magis compressa nunc angustata nunc valde dilatata:

V. badenensis D'ORB. Tab. XII, figg. 662—663.

Fig. 662: e mari Atlantico boreali; *a:* facies oralis.
Fig. 663: minor, e sinu Bergensi Norvegiæ; *a:* margo.

Vaginulina badenensis D'ORB, 1846, For. Bass. tert. Vienne, p. 65, t. 3, ff. 6—8.
        »   »   NEUGEB., 1856, For. Stichosteg. Lapugy, Wien. Akad. Dkschr. 12, 2, p. 98,
                        t. 5, ff. 7—9.
? Vag. marginata, caudata D'ORB., 1826, Tabl. méth., Ann. Sc. Nat. 7, p. 258, N:is 7—8.
? Vag. denudata v. REUSS, 1862, Nordd. Hils'n. Gault; Wien. Akad. S. Ber. 46, p. 45, t. 3, f. 4,
    et multa nomina alia.

Hab. mare Atlanticum boreale profund. metr. 1,740 haud frequens, long. mm. 4—5; ad Grip Norvegiæ
    (LILLJEBORG): long. mm. 2—3.

V. linearis MONTAGU.

Tab. XII, fig. 664.

Subteres vel paullum compressa, segmentis lineatis aut costulatis; sæpe seminuda.

Nautilus linearis MONTAG., 1808, Test. Brit. Supplem., p. 87, t. 30, f. 9.
Dental. legumen var. **linearis** WILLIAMS., 1858, Rec. For. Gr. Brit., p. 22, t. 2, ff. 46—48.
Vaginulina linearis BRADY, 1884, For. Chall. Rep. 9, p. 532, t. 67, ff. 11—12.
Synonym. ceteram vide Goës, l. c., p. 40.
A Nod. obliqua LIN. interdum difficillime distingvenda.

Hab. Skagerack præsertim ad Koster insulas metr. 50—180 passim. Long. mm. 5.

formæ affines:

1. plus minusve costata, paullum aut vix compressa, suturis sæpe impressis.

Inter Marg. costatam BATSCH et V. linearem MONTAG. medium tenens a neutra
certe limitanda, veris notis differentialibus destituta:

V. striatocostata v. REUSS, Tab. XII, fig. 665.

Fig. 665: facies lateralis et marginalis; *a:* facies oralis; e mari Caraibico.

Marginulina striatocostata v. REUSS, Nordd. Hils u. Gault, Wien. Ak. Sitz. Ber. 46, p. 62, t. 6,
    f. 2 (pygmæa).
? Vaginulina sulcata COSTA, 1855, Murna terz. Messina; Mem. Ac. Sc. Napol. 2, p. 145, t. 2, f. 17 (costis
    paucis).
Nodos. legumen var. **linearis** Goës, 1882, Retic. Rhizop. Carib. Sea; Sv. Vet. Akad. Handl. 19, 4, t. 2, f. 33.
Marginulina bononiensis FORNASINI, 1883, For. pliocen. Pontic. Savena; Boll. Soc. geol. Ital. 2, p. 187,
    t. 2, f. 7.

Hab. mare Caraibicum profund. metr. 500 rara (Goës).

2. saepe valde incrassata, aliquando angustata, plerumque paullum compressa, costis paucis plus minusve elatis, interdum obliquatis. Forma angustata a typica haud multum diversa.

V. costata BATSCH, Tab. XII, fig. 666.

Fig. 666: exemplum angustum, e mari Atlantico boreali; *a*: facies marginalis.

Nautilus costatus BATSCH, 1791, Conchyl. Seesands, t. 1, f. 1.
Marginulina Raphanus D'ORB., 1826, Tab. Méth.; Ann. Sc. Nat. 7, p. 258, t. 10, ff. 7—8; Mod. 6.
Marginulina costata BRADY, 1884, Chall: Rep. 9, p. 528, t. 65, ff. 10, 12—13.
        FORNASINI, 1893, For. Messin.; Mem. R. Accad. Sc. Istit. Bologna (5) 3, t. 2, f. 6.

Hab. mare Atlanticum boreale profund. metr. 1,740 rara (Jos. LINDAHL).

*** Nodosaria.

N. communis D'ORB.

In species male distinctas immerito divisa. Diversitas speciminum maxima fit: 1) magnitudine et constitutione segmenti embryonalis; quo maturius sive magis evolutum sit hoc segmentum, eo potius testa brevis segmentis paucis oritur; 2) segmentorum natura intumescendi maturorum, quorum suturæ constrictæ fiant; 3) incremento magnitudinis segmentorum plus minusve rapido; 4) septis inclinatis (obliquis) aut horizontaliter dispositis, nota tamen vacillans, septis segmentorum juvenilium obliquis, ceterisque transversis et vice versa interdum occurrentibus.

Ex quo formæ creantur maxime discrepantes, specifice tamen non distinctæ.

Ut species determinandi modum auctorum vitiosum perspicue videas, quasdam "species" modo eorum (id est notis vagantibus) describamus.

1. Nod. farcimen SOLD. transversim septata, extenuata, sensim crassitudine increscens, suturis modice constrictis, segmentis subinflatis; (forma modice increscens).

2. Nod. Boueana D'ORB. transversim septata, extenuata fere linearis, suturis vix aut paullum constrictis, segmentis elongatis ovatis; (forma lente increscens).

3. N. pauperata D'ORB. transversim septatata, crassior, brevior fere æqualis cylindrica, suturis segmentorum juvenilium non incisis, segmentorum suturis maturorum constrictis; (forma segmento embryonali majore, sensim increscens).

4. N. consobrina D'ORB. transversim septata, attenuata, præterea præcedenti similis.

5. N. consobrina var. emaciata REUSS longior, segmentis brevioribus magis numerosis.

6. N. plebeja REUSS transversim septata, cylindrica, partibus extremis ambabus plus minusve attenuatis, suturis non impressis (notæ cædem ac stadii juvenilis Nod. pauperatæ).

Siu vero notis talibus aliisque similibus utuntur auctores, differentia vera limitesque specifici extinguuntur; progenies fit species a parentibus disjuncta, delineatio artificialis non naturalis creatur. Varietates N. communis D'ORB. nominibus diversis conties relatas ad formas paucas plus minusve distinctas referre facile possumus.

Forma typica: Tab. XII, figg. 667—671.

Septis obliquis, sensim increscens linearis (Nod. badenensis D'ORB.) aut crassa, plus minusve raptim increscens, camera primordiali majori (N. Rœmeri NEUGEB., obliquata Rss, obliqua D'ORB. et aliæ, quæ in Marginulinam glabram D'ORB. et VAG. legumen LIN. transeunt.

    Figg. 667—668: e mari Azorico profund. metr. 530; a: facies oralis.
    Fig. 669: e fretis Koster insularum profund. metr. 100.
    Fig. 670: e mari Groenlandico profund. metr. 300—500.
    Fig. 671: megasphærica, e fretis Koster insularum profund. metr. 100.

Dentalina communis D'ORB., 1840, For. Craie bl. de Paris, Mém. Soc. géol. Fr. 4, p. 13, t. 1, f. 4.
    » inornata D'ORB., 1846, Bass. tert. Vienne, p. 44, t. 1, ff. 50—51.
Nodosaria communis BRADY, 1884, Challeng. Rep. 9, p. 504, t. 62, ff. 19—22.
    » GOÈS, 1882, Ret. Rhizop. Carib. Sea; Sv. Vet. Akad. Handl. 19, 4, t. 1, f. 16, t. 2, ff. 22, 24—25 (paullum compressa).
    FORNASINI, 1890, Lagen. plioc. Catanzarese; Mem. R. Accad. Sc. Istit. Bologna (4) 10, t. 1, ff. 14—18.

Hab. maria Groenlandica et Spetsbergica usque ad tropica rara.

formæ affines:

    1. nunc brevis, nunc elongata, septis horizontalibus, segmentis immaturis cylindricis, suturis obsolete aut haud incisis; segmentis maturis interdum plus minusve inflatis, ex quo accidit, ut formæ occurrant segmento primordiali magno, quæ segmenta omnia inflata habent, a N. soluta Rss vix distinctæ; aliæ breves, "Glandulinis" similes:

N. pauperata D'ORB., Tab. XII, figg. 672—688.

    Fig. 672: Dental. acuminata v. Rss; e mari arctico profund. metr. 380—890.
    Fig. 673: e mari Azorico profund. metr. 180.
    Figg. 674—676: e mari Groenlandico profund. metr. 380—530.
    Figg. 677—683: megasphæricæ, ex eodem loco.
    Fig. 684: procera, e mari Azorico profund. metr. 540.
    Fig. 685: angustior (= N. consobrina v. emaciata BRADY), e mari Caraibico profund. metr. 550.
    Fig. 686: forma eadem, e mari Azorico profund. metr. 890.
    Fig. 687: mesosphærica in N. solutam Rss transiens, N. farcimini SOLDANI, propinqua e mari Caraibico profund. metr. 500.
    Fig. 688: in Nod. solutam Rss transiens, e mari Groenlandico profund. metr. 430.

Dentalina pauperata D'ORB., 1846, Bass. tert. Vienne, p. 46, t. 1, ff. 57—58.
Nod. lævigata NILSS., 1827, Petrif. Suecana, p. 8, t. 9, f. 20.
? Naut. rectus MONTAG., 1803, Test. brit., p. 197, Suppl. 1808, p. 82, t. 19, ff. 4—7.
Dental. annulata v. Rss, 1851, Kreidemerg Lemberg; Haid. Nat. Abh. 4, p. 26, t. 1, f. 13; v. REUSS, 1872, Geinitz' Elbthalgeb, in Sachsen, 2, p. 85, t. 20, ff. 19—20.
    »   »   VAN DEN BROECK, 1876, For. Barbade; Ann. Soc. Belg. microscop. 2, t. 2, f. 2.
    »   »   FORNASINI, 1889, For. mioc. San Rufillo; Boll. Soc. geol. Ital. 5, t. 1, ff. 10—13.

Dental. communis var. pauperata PAHK., JONES & BRADY, 1866,.For. Crag., Palæogr. Soc. 19, p. 58,
    t. 1, f. 13—18, 20.
   communis, pauperata SHERB. & CHAPM., 1886, For. Lond. clay; Journ.·microsc. Soc. 1886,
    p. 750, t. 15, ff. 5—6, 9.
Nod. consobrina Rss, 1870, For. Sept. Thon. v. Pietzpuhl; v. SCHLICHT, t. 9, ff. 2, 8; t. 10, ff. 1,
    25—27.
 » acutioauda, vermioulum, bicuspidata, plebeja, Bötcheri REUSS, ibid. t. 8, f. 17; t. 9, ff. 10
    —12, 14, 16, 23; t. 10, ff. 5—12.
Dental. Verneuili D'ORB., 1846, Bass. tert. Vienne,.p. 46, t. 1, ff. 57—58.
Nod. fustiformis, tauricornis SCHWAG., 1866, For. Kar.-Nicobar, Novara Reise; geol. Theil 2, p. 228,
    t. 6, ff. 60—61 (forma procera).
 » consobrina var. emaciata Rss, (1851), 1870, v. SCHLICHT, t. 8, f. 15.
 »   »   » BRADY, 1884, Chall. Rep. 9, p. 502, t. 82, ff. 25—26.
 »   » BRADY, ibid. p. 501, t. 62, ff. 23—24 (in sequentem transiens).
   FORNASINI, 1890, Lagen. pliocen. Catanzaresc; Mem. R. Acc. Sc. Istit. Bologna (4) 10,
    t. 1, f. 12.

Quia Dent. consobrina D'ORB. est forma valde indistincta (fig. 3 D'ORB. Bass. tert. Vienne, t. 2, f. 3, valde
attenuata, a N. Boueana non distinguenda, et fig. 2 multo crassior a N. pauperata vix distincta), nomen "consobrinam" D'ORB. esse relinquendum censeo. Retinere nomen "consobrinam" longioribus et pauperatam brevioribus non est logicum et naturale. ·

Hab. mare Spetsbergense et Groenlandicum profund. metr. 180—500; mare Caraibicum passim. Long.
    mm. 5—12.

2. extenuata sensim increscens, septis horizontalibus, cameris elongatis ovalibus,
minus inflatis, aut subcylindricis. A N. ovicula D'ORB. vix nisi axi curvato
distincta; N. farcimen SILV. est forma talis recta vel subrecta:

N. Boueana D'ORB., Tab. XII, fig. 689.

Dent. Boueana, guttifera D'ORB., 1846, For. Bass. tert. Vienne, pp. 47, 49 t. 2, ff. 4—6,·11—12.
   consobrina (ex parte) D'ORB., 1846, ibid. p. 46, t. 2, f. 3 (segmento primordiali majore, quare
    testa magis æqualis, cylindrica oritur.
Nod. ovicula D'ORB., 1826, Tab. méth. An. Sc. Nat. 7, p. 252, No.·6.
? Naut. subarcuatus MONTAG., 1803, Test. Brit., p.·198, t. 6, f. 5.
Dent. peregrina REUSS, 1860, Crag. v. Antwerpen, Wien. Ak. S. Ber. 42, p. 356, t. 1, f. 6 (stadio
    juvenili suturis vix incisis, ut in præcedente).
    Lorneiana D'ORB., 1839, Craie blanche Paris; Mém. Soc. géol. France 4, p. 14, t. 1, ff.·8—9.
 » monile CORN., 1848, Nouv. foss. mier., Mém. Soc. géol. Fr. (2) 3, p. 250, t. 1, f. 18.
? Nod. Costai SCHWAG., 1866. For. Kar. Nicobar; Novara Reise, geol. Th. 2, p. 229, t. 6, f. 62.
? Dent. elegans D'ORB., 1846, Bass. tert. Vienne, p. 45, t. 1, ff. 52—56 (segmentis brevioribus).
 »   » SHERB. & CHAPM., 1886, For. Lond. Clay., Journ. mier. soc. 1886,·p. 750, t. 15, f. 7.
 » nepos, aduncs CORN., 1855, For. Marna bl. Vaticano, Mem. Accad. Sc. Napol. 2, pp. 1, 17,
    t. 1, ff. 1—2.
 » subtilis v. HREN, 1875, Clàv. Szab. Seh., p. 33, t. 3, f. 13.
Nod. gracilis v. REUSS 1845, Böhm. Kreidef. 1, p. 29, t. 8, f. 6.
Dent. trichostoma Rss, 1849, Neue For. Österr., Wien. Ak. Dkschr. 1, p. 367, t. 46, f. 6.
Nod. consobrina Rss, 1870, Sept. Th. Pietzp. v. SCHLICHT (partim) tab. 9, ff. 3, 22.
 » indifferens, Beningseni, approximata Rss, ibid. t. 2, ff. 6, 7, 13.
    farcimen SILV., 1872, Monogr. Nodos., Atti Accad. Gioen. Sc. nat. (3) 7, p. 83. t. 10, ff. 229—242.
   communis GOËS, 1882, Ret. Rhizop. Carib. Sea, Sv. Vet. Akad. Handl. 19, 4, t. 1, ff. 13—15.
 » filiformis BRADY (partim) 1884, For. Chall. Rep. 9, t. 63, ff. 3, 5.
    FORNASINI 1889, For. mioc. San Rufillo; Boll. Soc. geol. Ital. 5, t. 1, f. 14.
Dental. communis D'ORB., 1826 (= N. farcimen SOLD.), Tabl. méth., An. Sc. Nat. 7, p. 254, No.
    35, segmentis paullo brevioribus paullum inflatis, ub hac varietate merito dis-
    jungi vix potest. ·

Formæ paullo crassiores, suturis paullum constrictis, segmentis plus minusve inflatis, ad Nod. farcimen Rss, N. laxam Rss, Dent. distortam Costa etc. relatæ, inter hanc et Nod. solutam Rss medium tenent.

Hab. mare Caraibicum profund. metr. 300—500 (Goës).

N. soluta v. Reuss.

Tab. XII, fig. 690.

Septis transversis, segmentis inflatis fere globularibus, suturis plus minusve coarctatis.

Fig. 690: e mari Atlant. boreali profund. metr. 1,400.

Dental. soluta v. Reuss, 1851, For. Sept. Thon. Berlin, Zeitschr. deutsche geol. Gesellsch. 3, p. 60, t. 3, f. 4.
Nod. grandis 1865, For. deut. Sept. Thon., Wien. Ak. Dkschr. 25, p. 131, t. 1, ff. 26—28: Reuss 1870 in Schlichts tab. 8, ff. 13—14.
  »   capitata Rss, ibid. t. 8, f. 10.
  :   soluta Fornasini, 1890, Lag. pliocen. Catanzarese; Mem. R. Acc. Sc. Istit. Bologna (4) 10, f. 8.
  »   farcimen Brady, 1884, For. Chall. Rep. 9, t. 62, ff. 17—18, ab huc vix distincta.
  »   soluta Brady, ibid. p. 503, t. 62, ff. 13—16.

Hab. Atlanticum boreale et temperatum profund. metr. 1,400.   Long. mm. 5—6.

N. obliqua Lin.

Tab. XII, figg. 691—696, Tab. XIII, figg. 697—699.

Plerumque transversim septata, nunc recta vel subrecta, nunc, sæpissime, curvata; segmentis maturis sæpe inflatis, plus minusve lævibus, reliquis cylindricis, costatis vel lineatis, costis interdum interruptis et confluentibus.

Stadium juvenile costatum, maturum læve facit, ut forma obliqua ut typica et communis ut derivata jure majore sit habenda. Interdum subcompressa; etiam oblique septata.

Fig. 691: exemplum e mari Caraibico profund. metr. 500—600.
Fig. 692: exemplum e mari Atlantico boreali profund. metr. 1740. Dental. acutæ d'Orb. proxima.
Figg. 693—694: e mari Azorico profund. metr. 178—980; 694 in "vertebralem" vergens.
Fig. 695: e Gullmaren, sinu Bahusiæ profund. metr. 70.
Fig. 696: e sinu Skagerack profund. metr. 80; in obliquatam Batsch. vergens.
Fig. 697 (t. XIII): in "vertebralem" transiens, e mari Caraibico profund. metr. 500—600.
Figg. 698—699: vertebralis Batsch., Brady; conica Silv.; ex eodem loco.

Nautilus obliquus testa recto-subarcuata, articulis oblique striatis Lin., 1758, Syst. Nat. Ed. X, p. 710 (an = Naut. obliquatus Batsch).
Dental. subarcuata var. jugosa (ex parte) Williams, 1858, Rec. For. Gr. Brit., p. 20, t. 2, f. 42.

Nod. obliqua BRADY, 1884, Chall. Rep. 9, p. 513, t. ;64, ff. 20—22.
, , FORNASINI, 1892, Mem. Acc. Sc. Istit. Bologna (5) 2, ff. 1—7. Vide præterea Goës, 1882, Rhizop. Carib. Sea, Sv. Vet. Ak. Handl. 19, 4, p. 32.

Obs. N. vertebralis (BATSCH) BRADY, Chall. Rep., p. 514, t. 64, ff. 11—14; Goës, Rhizop. Carib. Sea, t. 1, f. 18, a N. obliqua non merito nomine speciali distingui potest.
Nota characteristica: septis crassis, quasi vitreis, non limbatis est minus stabilis quam quæ nomen distinctum legitimum faciat. Formæ sæpe occurrunt intermediæ inter *obliquam* typicam et illam. Crassitudo, altitudo et numerus costarum valde variantes. Sæpe recta a Nod. conica SILV. et N. Raphanistrum SILV. non distincta. Hæc est megasphærica illius forma.

Hab. ad oras Sveciæ rara metr. 30—120, Hardanger sinum Norveg. metr. 90—180 (NORMAN); mare Groenlandicum metr. 300—1,250, Spetsbergicum metr. 140, mare Atlant. boreale metr. 1,700 passim. Long. mm. 6—10.

forma affinis: seminuda septis transversis, stadio juvenili costato, aut lineato, ceterum lævis.

N. seminuda Rss, Tab. XIII, fig. 700.

Dental. seminuda 1849, Neue For. Österreichs; Wien. Ak. Dkschr. 1, p. 367, t. 46, f. 9.
Goës, Rhizop. Ret. Carib. Sea, Sv. Vet. Ak. Handl. 19, 4, p. 33, d) tab. 1, f. 17.

Hab. mare Caraibicum metr. 530. Long. mm. 18—22.

N. striolata Goës.

Tab. XIII, fig. 701.

Tenue striata, suturis stadii juvenilis vix incisis, segmentis ceteris subglobularibus, inflatis.
Ad Nod. solutam Rss correspondens. Goës, 1882, Ret. Rhizop. Carib. Sea; Sv. Vet. Ak. Handl. 19, 4. t. 1, f. 19.

Nod. soluta BRADY, 1884, Challeng. Rep. 9, t. 64, f. 28.
? **Dental. substriata** D'ORB., 1826, Tab. méth., An. Sc. Nat. 7, p. 255, No. 46.

Hab. ad Azores insulas metr. 200—600 (SMITT & LJUNGMAN); mare Caraibicum (Goës).

N. lævigata D'ORB.

Tab. XIII, figg. 702—703, 706—707, 709.

Ovoidea, ovoideo-globosa, ovoideo-apiculata, aut elipsoidea, haud raro septis primis obliquis, quare forma marginulinoidea exstat ("Psecadium" REUSS).
Fig. 702: exemplum e Skagerack profund. metr. 240—890.
Fig. 703: "psecadi"formis ex eodem loco.
Fig. 706: var. inflata BORNEMANN, v. REUSS, obtusissima v. REUSS; e mari Spetsbergensi profund. metr. 1,300.
Fig. 707: var. elliptica v. REUSS; e freto Koster insularum profund. metr. 180.
Fig. 709: septis obliquatis, e mari spetsbergensi profund. metr. 2,500.

Nod. (Glandulina) lævigata D'ORB., 1826, Tab. méth., Ann. Sc. Nat. 7, p. 252, t. 10, ff. 1—3.
Glandul. globulus, inflata, lævigata, obtusissima, rotundata v. REUSS, 1870; v. SCHLICHT, For. Sept.
            Thon Pietzpuhl, t. 6, ff. 1—3, 10—13, 17.
Nodosaria lævigata, rotundata BRADY, 1884, Challeng. Rep. 9, pp. 490—491, t. 61, ff. 17—18, 20—22.

Hab. ad oras Sveciæ et Norvegiæ occidentales, nec non in mari Arctico profund. metr. 20—2,500 haud rara.

forma affinis:

cylindrica aut obtuse fusiformis, plus minusve elongata, suturis aliquando plus minusve obliquatis, quare forma submarginulinoidea.

N. æqualis v. REUSS, Tab. XIII, figg. 704—705, 708, 710—711.

Fig. 704: pygmæa e Morup Sveciæ profund. metr. 18.
Fig. 705: pygmæa e mari Azorico, Marg. glabræ D'ORB. similis.
Fig. 708: subtypica e mari Atlantico boreali profund. metr. 500.
Fig. 710: .e mari Spetsbergensi profund. metr. 2,500.
Fig. 711: marginulinæformis, paullum compressa, e mari Atlantico boreali profund.
       metr. 1,740; a: facies marginalis.

Glandulina lævigata v. REUSS, 1863, Septar. Thon Offenbach, Wien Akad. S. Ber. 48, t. 3, f. 28.
      »   »   var. **æqualis** v. RSS, 1870, v. SCHLICHT, Septar. Thon Pietzpuhl, t. 6, ff. 21
                                      —22, 24.
Nod. lævigata var. **æqualis** BRADY, 1884, Challeng. Rep. 9, p. 492, t. 61, f. 32.
Glandulina æqualis FORNASINI, 1886; Gland. æqualis, Boll. Soc. geol. Ital. 5, t. 7, ff. 1—12.

Hab. cum præcedente minus frequens.

N. calomorpha v. RSS.

Tab. XIII, figg. 712—713.

Segmentis plus minusve inflatis, ovalibus aut ellipticis magnitudine fere æqua-
libus, paucis.

Species valde indistincta a N. radicula aut potius N. pauperata immerito disjuncta, cujus forma megasphærica pygmæa esse videtur.

RSS, 1865, For. deutsch. Septar. Thon; Wien. Ak. Dkschr. 25, p. 129, t. 1, ff. 15—19; 1870, v. SCHLICHT
        tab. 7, ff. 1—3.
Nod. consobrina PARK. & JONES, 1865, N. Atl. and Arct. Oc., Phil. Trans. 155, p. 342, t. 16, f. 3.
    »   **calomorpha** BRADY, 1884, For. Chall. Rep. 9, p. 497, t. 61, ff. 23—27.
? »   **pusilla** FORNASINI, 1890, Lagen. pliocen. Catanzarese, Mem. R. Acrad. Sc. Istitut. Bologna (4) 10, t. 1,
        ff. 9—10.

Obs. Nod. geinitziana NEUGB. eadem species ac N. radicula LIN. a BRADY est facta; illa tamen a N.
   calomorpha RSS differentiam specificam vix nullam exhibet.

Hab. in Bucken sinu Norvegico paucas invenit Cel. NORMAN metr. 260—350, pygmæas, vitreas, long. mm. 0.23—0.33.

N. scalaris BATSCH.

Tab. XIII, figg. 716—718.

Differentia inter N. scalarem et N. raphanum LIN. est multo minus distincta quam auctores vulgo concedant. Notæ characteristicæ scalaris: collum ore limbato productum, mucro segmenti embryonalis, segmenta testæ pauca ad summum VIII inflata non sunt sat validæ. Formæ in raphanum LIN. transeuntes sæpe occurrunt. Talis est Nod. raphanus GOËS, Rhiz. Caribb. Sea; Sv. Vet. Akad. Handl. 19, 4, t. 1, f. 9—10; et aliæ segmentis minus inflatis, fere cylindricis aut prismaticis.

Fig. 716: e mari Germanico profund. metr. 180.

Figg. 717—718: e mari Azorico profund. metr. 200.

Nod. scalaris D'ORB. medium inter N. raphanum et N. scalarem BATSCH tenet; conf. SILVESTRI l. c. t. 4, illa microsphærica hæc mesosphærica.

Nautilus BATSCH, 1791, Conchyl. d. Seesand. t. 2, f. 4.
? **Nod. Soldanii** D'ORB., 1826, Tabl. méth., Ann. Sc. nat. 7, p. 254, No. 30; **longicauda** ibid. No. 28.
N. longicauda SILV., 1872, Faun. mier. terr. subappen. Ital.; Atti Accad. Gioen. Sc. nat. (3) 7, p. 58, t. 5, ff. 101—106.
N. scalaris BRADY, 1884, For. Chall. Rep., p. 510, t. 63, ff. 28—31.
»   »   BRADY, PARK., JONES, 1887, Abrolhos Bank, Trans. zool. Soc. Lond. 12, 7, p. 223, t. 44, f. 6.

Hab. mare Germanicum metr. 100—178; in sinu Hardanger norvegico metr. 90—180 invenit NORMAN. Long. mm. 0.50—1.50.

APPENDIX.

N. carinata D'ORB.

Tab. XIII, figg. 714—715.

Plus minusve compressa, stichostegica aut stadio juvenili helicostegico (Lingulinopsis RSS). Apertura plerumque rimæformis.

Fig. 714: forma Lingulinopsis, e mari Azorico profund. metr. 450; a: facies oralis.

Fig. 715: pygmæa eodem ex loco; a: facies oralis.

* typica lata:

Lingulina carinata D'ORB., 1826, Tabl. méth., Ann. Sc. nat. 7, p. 257, No. 1, Mod. 26.
Frond. major, intumescens BORNEM., 1854, Liasform. Gölting, p. 36, t. 3, ff. 19—21.
Ling. carinata WILL., 1858, Rec. For. Gr. Brit., p. 14, t. 2, ff. 33—35.
Ling. ampullacea, polymorpha, mediterranea COSTA, 1861, Microdoride mediterranea 1, p. 45—47, t. 8, ff. 1—8.
Ling. decipiens STACHE, 1865, Tert. Merg. Whaingaroa Haf; Novara Reise Geol. 1, p. 193, t. 22, f. 17 (valde abbreviata).
Ling. Makowskyiana RZEHAK, 1885, Verhandl. Brünn. Nat. Verein. 24, p. 97, t. 1, f. 7.
Vide præterea GOËS, Ret. Rhizop. Caribb. Sea, Sv. Vet. Akad. Handl. 19, 4, p. 58.

** angusta:

Ling. carinata BRADY, 1884, Chall. Rep., p. 517, t. 65, ff. 16- 17.
» » SHERBORN & BAYLEY, 1890, Red. Chalk Foram., Journ. microsc. Soc. 1890, p. 558, t. 10, f. 3.
» » BALKW., MILLETT, 1884, For. Galway, Journ. micr. Sc. 3, t. 4, f. 6.
Vide prætcrea GOËS, Ret. Rhizop. Caribb. Sea, Sv. Vct. Akad. Haudl. 19, 4, p. 14.

*** stadio juvenili helicostegico:
Marginulina carinata D'ORB., 1826, Tab. méth., An. Sc. nat. 7, p. 259, No. 8.
Lingulinopsis bohemica Rss, Fiés' modell; No. 57.
» carlofortonsis BORNEM., 1884, Alti Soc. nat. Toscana 6, 1, p. 26, t. 6, ff. 1—7.

Hab. mare Azoricum, metr. profund. 500 (SMITT & LJUNGMAN).

LAGENA WALK. & BOYS.

L. lævis WALK. & BOYS.

Tab. XIII, figg. 719—722.

Fere globosa, aut ovoidea aut elliptica, collo plus minusve producto, »ectosolenica».
Fig. 719: globosa e mari Azorico profund. metr. 530; *a:* facies oralis.
Figg. 720—722: in variet. clavatam D'ORB. transiens; e mari Arctico.

Serpula lævis WALK. & BOYS., 1784, Test. min., p. 3, t. 1, f. 9.
Lag. lævis PARK. & JONES, 1857, For. Coast of Norway; A. M. Nat. Hist. (2) 19, p. 279, t. 11, f. 22.
Lag. vulgaris (typ.) WILLIAMS, 1858, Rec. For. Gr. Brit., p. 4, t. 1, fig. 5, 5 a.
Lag. vulgaris Rss, 1862, Lagenideen; Wien. Ak. Sitz. Ber. 46, p. 321, t. 1, fig. 15; t. 2, ff. 16—17; 1870,
v. SCHLICHT, For. Pietzpuhl, t. 2, ff. 3—8, 11.
Lag. suloata var. lævis PARK. & JONES, 1865, For. North Atl. and Arct. Or., Phil. Transact. 155, p. 349,
t. 13, f. 22, t. 16, f. 9 a.
Lag. antiqua ALCOCK,'1868, Life Hist. Foramf.; Mem. Manchest. Liter. & Philos. Soc. (3) 3, p. 176, t. 4, f. 3.
Lag. lævis BRADY, 1884, Chall. Rep. Zool. 9, p. 455, t. 56, ff. 14, 31.
Lag. bullæformis SCHWAG., 1867, Die Zone Ammonii. Sowerbyi; BENECKES Beiträge 1, 1868, p. 655, t. 34,
f. 5 (apiculata).

Hab. Skagerack profund. metr. 18—50; mare Groenlandicum et Spetsbergense profund. metr. 50—900 passim.
Long. mm. 0,60, rare ultra.

formæ affines:

1. Scabriuscula, hispido-aculeata, i. e. tubulis pseudopodialibus paullum porrectis
plus minusve horrida.

Tab. XIII, fig. 723: exemplum e sinu Bukken Norvegiæ.

Lag. hispida Rss.
(1858) 1862 Lagenideen; Wien. Ak. Sitz. Ber. 46, p. 335, t. 5, f. 66; t. 6, ff. 77—80.
Lag. oxystoma, hystrix Rss ibid.
Lag. hispida Rss, 1870, v. SCHLICHT, For. Pietzpubl, t. 3, ff. 26—27, t. 4, f. 4.
Lag. vulgaris var. oxystoma SHERBORN & CHAPMAN, 1886, Microz. Lond. Clay; Journ. Microsc. Soc.
1886, p. 744, t. 14, f. 15.
Lag. hispida BRADY, 1884, For. Chall. Rep., p. 459, t. 57, ff. 1—4.
Lag. Jeffreysi (BRADY) WRIGHT, Proc. Belf. Nat. Field Club 1876—77, Append. t. 4, f. 15.

Hab. in sinu Bukken profund. metr. 270—350 (Rev. NORMAN); e mari Spetsbergensi profund. metr.
2,315; exigua, quasi arenosa.

2. Testa tuberculato-spinosa, aut verrucosa.

Lag. tuberculata KARR., Tab. XIII, fig. 724.
1870, Kreideform. Leitzsdorf; Jhb. k. k. geol. Reichsanst. 20, p. 168, t. 10, f. 6, a.
L. aspera BRADY, vix diversa; For. Chall. Rep. p. 457, t. 57, ff. 7—12.
? L. aspera BALKW. & MILLETT, 1884, Journ. micr. Sc. 13, t. 2, f. 1.

Hab. Azores, profund. metr. 530 (SMITT & LJUNGMAN).

3. Clavato-lanciformis, interdum distoma.
L. clavata D'ORB., Tab. XIII, figg. 725—727.
Figg. 725—727: e mari arctico, in L. gracillimam SEG. transiens.

Oolina clavata D'ORB., 1846, For. Bass. tert. Vienne, p. 21, t. 1, ff. 2—3.
Amphorina elongata COSTA, 1854, Pal. Nap. 2, p. 122, t. 11, f. 12.
L. vulgaris v. olavata WILL., 1858, Rec. For. Gr. Brit., p. 5, t. 1, f. 6.
L. acicula Rss, 1860, For. Crag. v. Antwerp.; Wien. Ak. S. Ber. 42, p. 355, t. 1, f. 1.
L. olavata Rss, 1862, Lagenideen, Wien. Ak. S. Ber. 46, p. 320, t. 1, ff. 13—14.

Hab. Gullmaren Bahusiæ sinu metr. 70, Skagerak metr. 500, ad oras Groenlandiæ, Spetsbergiæ, metr. 40—200. Long. mm. 0.90.

4. Magis regulariter fusiformis, recta aut curvata, distoma; a præcedente male distincta.
L. gracillima SEGU. Tab. XIII, figg. 728—730.
Figg. 728—730: exempla e mari arctico.

Amphorina SEGU., 1862, For. monothal. mioc. Mess., p. 51, t. 1, f. 37.
Amphorina gracilis COSTA, 1854, Pal. Nap. 2, p. 121, t. 11, f. 11.
L. suloata var. distoma polita PARK. & JONES, 1865, N. Atl. and Arct. Oc., Phil. Transact. 155, p. 357, t. 13, f. 21.
L. gracillima PARK., JONES, BRADY, 1866, Crag. For., Pal. Soc. 19, p. 45, t. 1, ff. 36—37.
L. vulgaris v. distoma-polita RYM., JONES, 1872, Java deep Sea Lagen, Lin. Soc. Trans. 30, p. 64, t. 19, f. 55.
L. gracillima BRADY, 1884, For. Chall. Rep., p. 456, t. 56, ff. 19—28.

Hab. mare Spetsbergiæ metr. 900. Long. 0.50—1.50.

5. Magis cylindrica, distoma; interdum scabra, quasi arenacea; a præcedente non sat distincta.
L. elongata (EHRENB.) BRADY. Tab. XIII, fig. 731.
Fig. 731: e mari Groenlandico profund. metr. 40.

1884, For. Chall. Rep., p. 457, t. 56, f. 29.

Hab. mare Groenland. & Spetsberg. cum præcedente profund. metr. 40—2,300. Long. mm. 0.60—1.

L. striata D'ORB.
Tab. XIII, figg. 732—736.

Typi habitu, costis aut lineis longitudinaliter ornata, collum sæpe annulis vel lamina spirali præditum; fundus interdum crenatus aut spinescens; aliquando bicamerata.
Figg. 732, 734—735: exempla e Gullmaren sinu Bahusiæ profund. metr. 35—100.
Fig. 133: forma bicamerata R. JONES, eodem ex loco.
Fig. 736: costata (V. perlucidum MONTAG), e mari Azorico profund. metr. 530.

* striata.

Oolina striata D'ORB., 1839, Voy. Amér. mér., p. 21, t. 5, f. 12.
Ool. Haidingeri CZJZ., 1847, For. Wiener. Beck.; Haid. nat. Wiss. Abh. 2, p. 138, t. 12, ff. 1—2.
L. vulgaris v. substriata WILL., 1858, Rec. For. Gr. Brit., p. 7, f. 14.
L. striata, gracilicosta (tubulifera, Haidingeri) Rss, 1862, Lagenid., Wien. Ak. S. Ber. 46, p. 326—327,
        t. 3, ff. 41—45; t. 4, ff. 46—47; 1863, Crag. d'Anvers; Bull. Ac. Belg. (2)
        15, p. 142, t. 1, ff. 10—11 (seminuda).
L. tenuistriata STACHE, 1865, Tert. Mergel Whaingaroa Haf., Nov. Reise. geol. 1, p. 184, t. 22, f. 4.
? L. grinsingensis KARR., 1877, Hochquellen Wasserleit.; K. K. geol. Reichsanst. Abh. 9, p. 378, t. 16, f. 17.
L. striata Rss, 1870, Sept. Thon Pietzpuhl, Wien. Ak. S. Ber. 62, p. 467; v. SCHLICHT, t. 8, t. 1, ff. 7—11.
L. striata BRADY, 1884, For. Chall. Rep. 9, p. 460, t. 57, ff. 19, 22, 24, 28.
L. striata BRADY, PARK. & JONES, 1887, For. Abrohlos Bank, Lond. Zool. Soc. Transact. 12, 7, p. 222, t. 44, f. 28.

** costata, a præcedente non distincta; interdum apiculata aut distoma.

Vermiculum perlucidum MONTAG, 1803, Test. Brit., p. 525, t. 14, f. 3.
L. perlucida, striata, interrupta (partim) WILL., 1858, Rec. For. Gr. Brit., p. 5, 6, 7, ff. 8, 10, 11.
? L. bifrons GUMB. (1868) 1870, For. Nordalp. Eocän, Münch. Akad. Wiss. 10, p. 607, t. 1, f. 9.
L. filicosta, amphora Rss, 1862, Lagenid., Wien. Ak. S. Ber. 46, p. 328, t. 4, ff. 50—51; 57, Crag d'Anvers;
        Bull. Ac. Belge (2) 15, p. 143, t. 1, f. 12.
L. sulcata PARK., JONES, 1865, N. Atl. Arct. Oc., Phil. Trans. 155, t. 13, f. 28.
L. Lyellii (Segm. 1862) BRADY, 1871, Brackish Water For., A. N. H. (4) 6, p. 292, t. 11, f. 7 (distoma?)
L. sulcata WRIGHT, 1877, Belf. nat. Field Club. Proceed., 1876—77, Append., t. 4, f. 10.
L. sulcata BRADY (partim), 1883, For. Chall. Rep., t. 57, ff. 23, 26; t. 58, ff. 4, 17.
L. sulcata v. interrupta ibid., t. 57, ff. 25, 27; t. 58, ff. 5—6.
PARK., JONES, BRADY, 1887, For. Abrohlos Bank, Trans. Lond. Zool. Soc. 12, 7, p. 222, t. 44, ff. 18, 22, 34.
Lag. Lyellii BALKW. & MILLETT, 1884, Journ. micr. & nat. Sc. 3, t. 2, f. 2.

Hab. ad oras Scandinaviæ metr. 50—350, passim.   Long. mm. 0.40—0.80.

formæ affines:

    1. forma typicæ aut clavæformis, fundo striato aut costato.

L. semistriata WILL. Tab. XIII, figg. 737.

L. striata v. semistriata WILL., 1848, Monagr. Lagenid. Gr. Br.; A. N. H. (2) 1, p. 14, t. 1, ff. 9—10.

* fundo striato:

Oolina striaticollis D'ORB., 1839, For. Amér. mér., p. 21, t. 5, f. 14.
Ovulina tenuis (partim.), laoryma BORNEMANN, 1855, Sept. Thon v. Hermsdorf; Zeitschr. deut. geol.
        Gesellsch. 7, p. 317, t. 12, ff. 2—3.
Ool. punctata, striatula EGGER, 1857, Mioc. Ortenburg, Leonh. u. Bronns Jhb. 1856, p. 268, 269,
        t. 5, ff. 1—8.
L. vulgaris v. perlucida (partim), WILL., 1858, Rec. For. Gr. Brit., p. 5, 6, t. 1, f. 7.
L. vulgaris v. semistriata, tenuis Rss, 1862, Lagenideen; Wien. Ak. S. Ber. 46, pp. 322, 325, t. 2,
        ff. 18—21, t. 3, ff. 33—39; REUSS, 1863, Crag d'Anvers, Bull. Ac. Belg.
        (2) 15, p. 141, t. 1, ff. 3—9; v. SCHLICHT, For. Pietzpuhl, t. 3, f. 12.
L. semillineata WRIGHT, 1885, Proc. Belf. nat. Field Club, Append. 1885—86, p. 320, t. 26, f. 7.

** fundo costato:

L. vulgaris v. semistriata WILL., Rec. For. Gr. Brit., p. 6, t. 1, f. 9.
L. sulcata v. semistriata PARK. & JONES, 1865, N. Atl. und Arct. Oc. Phil. Trans. 155, t. 13, f. 23.
L. tenuis REUSS, 1870, v. SCHLICHT, Foram. Pietzpuhl, t. 2, ff. 12—18, 23.
L. semistriata BRADY, 1884, For. Chall. Rep., p. 465, t. 57, ff. 14, 16—17.
    »   »   WRIGHT, 1886, Proc. Belf. nat. Field Club Append. 1885—86, t. 26, f. 6.

Hab. Oceanum boreale rara, pygmæa, profund. metr. 1,700; in Skagerrek sinu profund. metr. 700 (Jos.
    LINDAHL).

2. elliptico-clavæformis aut lanciformis, striata vel costata var. L. clavatœ correspondens.

L. gracilis WILL., Tab. XIII, fig. 738.

L. vulgaris v. gracilis WILL., 1858, Rec. For. Gr. Brit., p. 7, t. 1, ff. 12—13.
Rss, 1862 (partim) Lagenid., Wien. Ak. S. Ber. 46, p. 331, t. 4, ff. 58—61; t. 5, f. 62; 1870, For. Sept. Thon. Pietzp. Wien. Ak. S. Ber. 62, p. 467, v. SCHLICHT, tab. 2, ff. 19—20, 24—25.
L. sulcata PARK. & JONES (partim), 1865, N. Atl. & Arct. Oc. Phil. Trans. 155, p. 351, t. 16, f. 7.
L. gracilis BRADY, 1884, For. Chall. Rep., p. 464, t. 58, ff. 2, 3, 7—10, 22—24 (aluto-costata).

Hab. Skagerack metr. 530—890 (Théel, Bowallius); mare glaciale Spetsb. metr. 1,790. Long. mm. 0.25—0.50.

3. fusiformis, distoma, tenuissime, interdum obsolete striata.

L. distoma (PARK. & JONES) BRADY. Tab. XIII, figg. 739—740.

1864, Rhiz. Shetland., Trans. Lin. Soc. 24, p. 467, t. 48, f. 6.
? L. lævis v. striata PARK. & JONES, 1857, For. Coast of Norway, A. N. H. (2) 19, p. 279, t. 11, f. 24.
L. sulcata v. distoma PARK. & JONES, 1865, N. Atl. & Arct. Oc., Phil. Trans. 155, p. 356, t. 13, f. 20.

Hab. Skagerack metr. 60; mare arct. Spetsbergiæ & Groenlandiæ metr. 350—1,700. Long. mm. 0.50—2.00.

L. Hertwigiana.

Plerumque pyriformis, sub-lævis, collo plerumque brevi, testa interne lacunosa, lacunis singulis ostio poriformi exeuntibus; ostia interdum in series disposita.

Lag. hertwigiana BRADY, 1884, Chall. Rep. Zool. 9, p. 470, t. 58, f. 36.

Hab. in Bukken sinu Norvegico profund. metr. 156—260, specimina pauca emacinta collegit Rev. NORMAN.

L. globosa WALK. & JAC.

Tab. XIII, fig. 741.

Globosa, ovo-pyriformis aut elliptica, interdum anguste ovalis fere cylindrica; collo retracto (= Entosolenia) aut valde abbreviato, fundo interdum valde spinoso.

Forma ovalis (= Miliola ovum EHRENB. Microgeol. t. 27, f. 1 &c; Lag. ovum BRADY, For. Chall. Rep., t. 56, f. 5) non sat distincta.

Serpula globosa WALK. & BOYS, 1784, Test. micr., p. 3, t. 1, f. 8.
Oolina inornata D'ORB., 1839, For. Amér. mérid., p. 21, t. 5, f. 13.
Lag. inornata Rss, 1862, Lagenid., Wien. Ak. S. Ber. 46, p. 320, t. 1, f. 12.
Cenchridium oliva EHRENB., 1854, Microgeol., t. 24, ff. 3—4 (ovalis).
Lag. perovalis GÜMB., 1868, For. Nordalp. Eocän, Münch. Ak. Wiss. Abh. 10, 2, p. 606, t. 1, f. 7 (ovalis).
L. ovum MARSS, 1878, Schreibkreide Rügen; Mittheil. Greifswald. Nat. Verein 1878, 10, p. 120, t. 1, f. 1 (ovalis).
Phialina oviformis COSTA, 1854, Pal. Nap. 2, p. 123, t. 11, ff. 8—9.
Entosolenia globosa WILL., 1858, Rec. For. Gr. Brit., p. 8, t. 1, ff. 15—16.
Ool. simplex Rss, 1850, Kreidemerg. Lemberg; Haid. Nat. Wiss. Abh. 4, p. 22, t. 1, f. 2.
L. globosa Rss, 1862, Lagenid. Wien. Ak. S. Ber. 46, p. 318, t. 1, ff. 1—3; 1863, Crag d'Anvers, Bull. Ac. Belge (2) 15, p. 143, t. 1, ff. 13—14.

L. sulcata v. **globosa** PARK. & JONES, 1865, N. Atl. & Arct. Oc., Phil. Trans. 155, p. 348, t. 13, f. 37; t. 16, f. 10.
L. globosa Rss, 1870, For. Sept. Thon. Pietzpuhl; Wien. Ak. S. Ber. 62, p. 465--66; v. SCHLICHT, tab. 1, ff. 5—10.
BRADY, 1884, For. Chall. Rep. 9, p. 452, t. 56, ff. 1—4 (fig. 1 fere ovalis).
SHERB. & CHAPMAN, 1886, London Clay; Journ. Micr. Soc. 1886, p. 8, t. 14, f. 11; Red Chalk of Yorkshire etc. ibidem 1890, p. 555, t. 9, ff. 2, 4.

Hab. mare glaciale Spetsbergense metr 1,250—2,400 non frequens. Long. mm. 0.80—1.

APPENDIX.

var. hispida:
? **Lag. hystrix** Rss, (1858) 1862, Lagenid., Wien. Ak. S. Ber. 46, p. 335, t. 6, f. 80.
v. SCHLICHT, Septar. Thon. Pietzpuhl, t. 3, f. 28.
Ool. salentina COSTA, 1854, Pal. Nap. 2, p. 118, t. 11, ff. 13—14.
Lag. hispida WRIGHT, 1877, Rec. For. Down & Antrim; Proc. Belfast nat. Field Club 1876—77, Append. t. 4, f. 7.
L. aspera Rss, 1861, Kreidetuff v. Maastrich, Wien. Ak. S. Ber. 44, p. 305, t. 1, f. 5.
L. aspera, rudis Rss, 1862, Lagenid., Wien. Ak. S. Ber. 46, p. 335, 336, t. 6, ff. 81, 82.
REUSS, Crag d'Anvers, 1863, Bull. Ac. Belge (2) 15, p. 145, t. 1, f. 17.
L. Parkeriana BRADY, 1876, Carb. & Perm. For., Pal. Soc. 30, p. 120, t. 8, ff. 1—5.
Entosolenia rudis MOEB., 1880, Mauritius, p. 90, t. 8, f. 10.
? **L. vulgaris ampulla distoma** RYM., JONES, 1872, Java Deep Sea Lagen., Trans. Lin. Soc. 30, p. 63, t. 19, f. 52.
? **L. ampulla distoma** BRADY, 1884, For. Chall. Rep. 9, t. 57, f. 5.

L. sulcata WALK. & BOYS.

Tab. XIII, figg. 742—744.

Globosa aut oviformis nunc striolata nunc costata, plerumque entosolenica aut collo brevi, fundo aliquando perforato aut area circulari laevi instructo.

Fig. 742: L. lineata WILLIAMS, e mari Groenlandico; a: facies oralis.
Fig. 743: sulcata auctorum, e mari arctico; a, b: facies orales duae.
Fig. 744: minor, e Gullmaren sinu Bahusiae; a: facies oralis.

* striolata:
Entosolenia globosa var. **lineata** WILLIAMS, 1858, Rec. For. Gr. Brit, p. 9, t. 1, f. 17.
L. lineata Rss, 1862, Lagenid., Wien. Ak. S. Ber. 46, p. 328, t. 4, f. 48.
BRADY, 1884, For. Chall. Rep., p. 461, t. 57, f. 13.
? **L. striata** (partim) BRADY, ibid., t. 57, f. 30.
Miliol. striata EHRENB., 1854, Microgeol., t. 24, f. 5.
Entosolenia costata WILL., 1858, Rec. For. Gr. Brit., p. 9, t. 1, f. 18.
L. costata Rss, 1862, Lagen., Wien. Ak. Sitz. Ber. 46, p. 329, t. 4, f. 54.
L. Villardeboana Rss, 1863, Crag d'Anvers, Bull. Ac. Belg. (2) 15, p. 144, t. 1, f. 15.
Capitellina multistriata MARSSON, 1878, Rügen. Schreibkr.; Greifswald. Nat. Verein. Mittheil. 1878, p. 123, t. 1, f. 3.
Lag. costata WRIGHT, 1877, Rec. For. Down & Antrim, Proc. Belfast nat. Field Club 1876—77, Append. t. 4, f. 12.

** sulcata:
Vermiculum sulcatum WALK. & BOYS, 1784, Test. min., p. 2, t. 1, f. 6.
L. costata BALKW. & WRIGHT, 1885, Rec. Dublin For.; Trans. R. Ir. Ac. Sc. 28, p. 338, t. 14, ff. 3—5.
O. Villardeboana, Isabellae, raricosta D'ORB., 1839, Voy. Am. mérid., t. 5, ff. 4—5, 7—8, 10—11.

L. **Villardeboana, Isabellæ** Rss, 1862, Lagenid., Wien. Ak. S. Ber. 46, p. 329, t. 4, ff. 53, 55—56; v. SCHLICHT, t. 3, f. 19.
L. **mucronulata** Rss, ibid., p. 329, t. 4, f. 52; Rss, 1870, v. Schlichts tab. 3, f. 24.
Ovul. **elegantissima** BORNEM., 1855, Sept. Thou. Hermsdorf; Zeitschr. deutsch. geol. Gesellsch. 7, p. 316, t. 12, f. 1.
L. **sulcata** PARK. & JONES, 1865, N. Atl. & Arct. Oe., Phil. Trans. 155, p. 351, t. 13, ff. 28—30.
BRADY, 1884, For. Chall. Rep., p. 462, t. 57, ff. 33—34.
SHERBORN & CHAPMAN, 1886, London Clay, Journ. Micr. Soc. 1886, p. 745, t. 14, f. 18.
L. **acuticosta, subalata** Rss, 1861, Kreidetuff Maastrich; Wien. Ak. S. Ber. 44, p. 305, ff. 1, f. 4: 1863, Lagenid., Wien. Ak. S. Ber. 46, p. 331, t. 5, f. 63.
REUSS, 1870, Sept. Thon. Pietzpuhl, Wien. Ak. S. Ber. 62, p. 467, v. SCHLICHTS, tab. 3, f. 23.
BRADY, 1884, For. Chall. Rep., p. 464, t. 57, ff. 31—32.
BRADY, PARK., JONES, 1887, For. Abrohlos Bank, Trans. Lond. Zool. Soc. 12, p. 222, t. 44, ff. 26, 31.
L. **costifera** TERQUEM, 1886, For. et Ostrac. de l'Island, Bull. Soc. Zool. France, 1886, p. 330, t. 11, ff. 3, 4.

Hab. Skagerack, mare glaciale metr. 50—1,800 passim. Long. mm. 0.40—0.70.

formæ affines:

1. sulcata, basi colli arcolata sive reticulata. Forma inter præcedentem et sequentem; a quibus vix discernenda.

L. **Williamsoni** ALCOCK, 1865.

BALKW. & WRIGHT, 1885, Rec. Dubl. For., Proc. R. Irish. Ac. Sc. 28, p. 339, t. 14, ff. 6—8.
WRIGHT, 1877, Rec. For. Down & Antrim; Proceed. Belfast nat. Field Club 1876—77, Append. t. 4, f. 14.

Hab. ad Spetsbergiam metr. 180; in Stoksund metr. 150—180 reperit Rev. Norman; habitus et magnitudo præcedentis.

2. globosa aut pyriformis, sulcata, irregulariter areolata, areolis squamiformibus, aut costis transversis, rectis vel curvatis, magis regulariter areolata; areolis magnitudine valde variantibus.
Aliquando seminuda, aut costis transversis dimidia parte destituta.

L. **squamosa** MONTAG, Tab. XIII, 745.

Fig. 745: e mari Spetsbergico profund. metr. 180; a: facies oralis.

Vermiculum squamosum MONTAG, 1803, Test. Brit., p. 526, t. 14, f. 2.
? **Entosolenia squamosa** WILLIAMS, 1858, Rec. Brit. For., p. 12, f. 29 (areolis fere hexagonis, a sequenti male distincta).
Oolina melo D'ORB., 1839, Voy. Amér. mériod., p. 20, t. 5, f. 9.
Lag. globosa var. **catenulata** PARK. & JONES, 1857, For. Coast of Norway; A. N. H. (2) 19, p. 278, t. 11, f. 26.
Entosolenia squamosa v. **catenulata** WILL., 1858, Rec. For. Gr. Brit., p. 13, t. 1, f. 31.
L. **catenulata** Rss, 1862, Lagenid. Wien. Ak. S. Ber. 46, p. 332, t. 6, ff. 75—76.
L. **sulcata** v. **melo** PARK. & JONES, 1865, N. Atl. & Arct. Oc. Phil. Trans. 155, p. 353, t. 13, ff. 33—36.
L. **squamosa** BRADY, 1884, For. Chall. Rep. 9, p. 471, t. 58, ff. 28—31.
L. **melo** BRADY, PARK. & JONES, 1888, For. Abrohlos Bank, Trans. Zool. Soc. London 12, 7, p. 222, t. 44, ff. 21, 24 (25, sulcata-semiareolata).

Hab. Gullmaren sinum Bahusiæ metr. 150 (WIRÉN & AURIWILLIUS), mare Spetsbergiæ metr. 180 inter varietatem præcedentem; Bukken Norvegiæ metr. 260—350 (NORMAN).

3. globosa aut ovoidea, areolis plus minusve distincte hexagonis, interdum in series dispositis, ornata; areolis magnitudine variantibus.
Interdum seminuda.

L. hexagona WILL., Tab. XIII, fig. 746.

Fig. 746: e Farsund Norwegiæ profund. metr. 60.

Entosolenia squamosa var. hexagona WILL., Rec. For. Gr. Brit., p. 13, t. 1, f. 32.
Entosol. scalariformis WILL., l. cit., p. 13, f. 30.
L. reticulata, scalariformis (partim), favosa, geometrica Rss, 1862, Lagenid., Wien. Ak. S. Ber. 46,
            ff. 333, 334, t. 5, ff. 67—70, 72—73.
L. sulcata v. squamosa PARK. & JONES, 1865, N. Atl. & Arct. Oc., Phil. Trans. 354, t. 13, ff. 40—41,
            t. 16, f. 11.
L. hexagona BRADY, 1884, For. Chall. Rep. 9, p. 472, t. 58, f. 33.
   »   »   BALKWILL & MILLETT, 1884, For. Galway; Journ. micr. Sc. 3, t. 1, f. 10.

Hab. ad oras Bahusiæ & Norwegiæ metr. 50—350 passim, inter L. sulcatam; inter minimas. Long.
    mm. 0.30—0.50.

L. apiculata Rss.

Tab. XIII, fig. 747.

Plus minusve anguste elliptica aut pyriformis sæpe apiculata, interdum subcylindrica.
A L. globosa WALK. non sat distincta; nomen REUSSI ineptum.

Fig. 747: exemplum e mari Spetsberg.; a: facies oralis; b, c: alia exempla eodem
ex loco.

* non apiculata:

Oolina ellipsoides COSTA (ex parte), 1854, Pal. Nap. 2, t. 11, f. 25.
Cenchridium dactylus EHRENB., 1854, Microgeol., t. 24, ff. 1—2.
? Lag. Bittneri KARR., 1877, Hochquellen Wasserleit., Abh. K. K. geol. Reichsanst. 9, p. 378, t. 16, f. 18.
L. emaciata Rss, 1862, Lagen., Wien. Ak. S. Ber. 46, p. 319, t. 1, f. 9.
SHERB. & CHAPM., 1890, Red Chalk of Yorksh., Journ. Micr. Soc. 1890, p. 555, t. 9, ff. 8, 12—13.

** apiculata:

Oolina apiculata Rss, 1850, Kreidemerg. Lemberg; Haid. Nat. Wiss. Abh. 4, p. 22, t. 1, f. 1.
L. apiculata Rss, 1862, Lagenid., Wien. Ak. S. Ber. 46, p. 318, t. 1, ff. 4—8, 10—11.
var. elliptica Rss, 1862, Nordd. Hils und Gault, Wien. Ak. S. Ber. 46, p. 35, t. 2, f. 2.
L. (Entosolenia) sulcata var. apiculata PARK. & JONES, 1865, N. Atl. & Arct. Oc., Phil. Trans. 155, p. 358,
            t. 13, ff. 38—39.
PARK., JONES, BRADY, 1866, Crag. For. Pal. Soc. 19, p. 44, t. 1, f. 27.
Rss, 1870, For. Sept. Thon. Pietzpuhl Wien. Ak. S. Ber. 62, p. 466, v. SCHLICHT, t. 1, ff. 11—12, 14—18,
            20, t. 2, f. 1.
v. HANTK., 1875, Clav. Szaboi Schicht (separ.), p. 22, t. 12, ff. 7—9.
BRADY, 1884, For. Chall. Rep., p. 453, t. 56, ff. 15—18.
SHERBORN, CHAPMAN, 1886, Lond. Clay., Journ. Micr. Soc. 1886, p. 744, t. 14, f. 14; Red. Chalk of Yorksh.
    ibid., 1890, p. 555, t. 9, ff. 6—7, 9—11.

Hab. mare Spetsbergense metr. 1,780—4,600 sat frequens. Long. mm. 0.30—60.

formæ affines:
   1. »apiculata» semistriata.

L. semiornata TERQU., 1886, For. Ostrac. de l'Islande; Bull. Soc. Zool. France 1886, p. 330, t. 1, f. 2.

Hab. mare boreale (TERQUEM.)

APPENDIX.

2. Anguste pyriformis aut ellipsoidea, sæpe apiculata, striata.

L. caudata D'Orb.

Oolina caudata D'Orb., 1839, For. Voy. amér. mérid., p. 19, t. 5, f. 6.
Lag. caudata Rss, 1862, Lagenid., Wien. Ak. S. Ber. 46, p. 325, t. 3, f. 29.

L. marginata Walk. & Boys.

Tab. XIII, figg. 748—751.

Sublenticularis, pyriformis aut ovalis, plus minusve compressa, sæpe marginata-carinata aut alata, apertura rimæformis; fundo interdum spinoso; aliquando trigona.
Figg. 748—751: facies lateralis.
Figg. a, a: facies oralis formarum amborum.
Est Lag. globosa compressa, ab illa interdum difficillime distingvenda. Carina marginis nota non sat characteristica nec stabilis; fundo interdum subalato-crenato L. fimbriata Brady, Millett, For. Galway, Journ. microsc. Sc. 3, p. 82, t. 2, f. 5 (ala quasi duplicata) non carinata, haud multum compressa:

Fissurina lævigata Rss, 1849, Neue For. Österr. Tert. Beck., Wien. Ak. Denkschr. 1, p. 366, t. 46, f. 1;
1862, Lagenid., Wien. Ak. S. Ber. 46, p. 338, t. 6, f. 84; 1870, Sept. Thon.
Pietzpuhl, Wien. Ak. S. Ber. 62, p. 470; v. Schlicht, t. 4, ff. 16—24.
Fiss. globosa Reuss, v. Schlicht, t. 5, ff. 4—6.
Fiss. globosa Bornem., 1855, Sept. Thon. Hermsdorf; Zeitschr. deutsch. geol. Gesellsch. 7, p. 317, t. 12, f. 4
(fere globosa).
Fiss. obtusa Egger, 1857, Mioc. Ortenburg; Leonh. & Bronns Jhb. 1857, p. 270, t. 5, ff. 16—19 (fere globosa).
    »   Reuss, 1862, Lagenid., Wien. Ak. S. Ber. 46, p. 339—40, t. 5, ff. 4—6; t. 7, ff. 88, 92—93.
L. lævigata Brady, 1884, For. Chall. Rep. Zool. 9, p. 473, t. 114, f. 8.
    »   »   Balkw. & Mill., 1884, For. Galway, Journ. Micr. Sc. 3, t. 2, f. 6 (valde elongata); t. 3, f. 6
(trigona).
L. faba Balkw. & Mill., ibid. t. 2, f. 10 (non sat distincta).

** plus minusve carinata:

Interdum plano- aut concavo convexa (= var. inæquilateralis Wright).

Serpula (Lagena) marginata Walk. & Boys, 1784, Test. min., p. 2, t. 1, f. 7.
Ool. compressa D'Orb., 1839, For. Amér. mérid., p. 18, t. 5, ff. 1—2.
? Amygdalina calabra Costa, 1854, Pal. Nap. 2, p. 120, t. 11, f. 7.
Fiss. alata Reuss, 1851, Sept. Thon. Hermsdorf., Ztschr. deutsch. geol. Gesellsch. 3, p. 58, t. 3, f. 1; 1870,
Wien. Ak. S. Ber. 62, p. 469, v. Schlicht, t. 4, ff. 7—9, 13—15; t. 5,
ff. 19—21.
Entosolenia marginata (partim) Will., 1858, Rec. For. Gr. Brit., p. 10, f. 21.
Fiss. carinata, alata Rss, 1862, Lagenid., Wien. Ak. S. Ber. 46, p. 338, 389, t. 6, f. 83; t. 7, ff. 86—87;
v. Schlicht, t. 5, ff. 1—3.
Lag. sulcata var. marginata Park. & Jones, 1865, N. Atl. & Arct. Oc., Phil. Trans. 155, p. 355, t. 13,
ff. 42—43 (44?); t. 16, f. 12.
Lag. marginata Wright, 1891, Rec. For. S. Donegal, Proc. Belf. nat. Field Cl. 1880—81, Append., t. 8, f. 4.

L. marginata BRADY, 1884, For. Chall. Rep., p. 476, t. 59, ff. 21—23.
BRADY, PARK. & JONES, 1888, For. Abrohlos Bank; Trans. Zool. Soc. London 12, 7, p. 222, t. 44, ff. 27, 29, 30, 32.
L. marginata BALKW. & MILL., 1884, For. Galway, Journ. micr. Sc. 3, t. 3, f. 2.
      »   »   var. **inæquilateralis** WRIGHT, 1885, For. Down & Antrim; Proc. Belf. nat. Field Club 1884
             —85, t. 26, f. 10.
L. trigono-elliptica BALKW. & MILL., 1884, For. Galway; Journ. micr. Sc. 3, p. 13, t. 3, f. 8 (trigona).

Hab. ad oras Bahusiæ profund. metr. 20—150 passim; in sinu Codano metr. 160; mare Spetsbergicum metr.
    2,500. Diam. mm. 0.16—0.50.

formæ affines:

1. carinata aut alata, carina triplicata; interdum umbonata (= var. Walleriana
   WRIGHT, Rec. For. Ireland., Trans. R. Ir. Ac. Sc. (3) No. 4, p. 481, t. 20, f. 8).

L. orbignyana SEG.
Fissurina orbignyana SEG., 1862, For. monothal. Messina, p. 66, t. 2, ff. 25—26.
? **Ool. compressa** D'ORB., 1846, Bass. tert. Vienne, p. 23, t. 21, ff. 1—2.
Entosolenia marginata (partim) WILL., 1858, Rec. For. Gr. Brit., p. 110, ff. 19—20.
Lag. marginata Rss, 1862, Lagenid., Wien. Ak. S. Ber. 46, p. 222, t. 2, ff. 22—23.
? **Lag. tricincta** GÜMB., 1868, For. Nortalp. Eocän; Abh. Bay. Ak. Wissensch. 10, p. 606, t. 1, f. 8.
Fiss. carinata Rss, 1870, For. Sept. Thon. Pietzpuhl; Wien. Ak. S. Ber. 62, p. 469, v. Schlicht,
    t. 5, ff. 13—15.
Lag. carinata var. **orbignyana** WRIGHT, 1881, For. S. Donegal, Proc. Belf. nat. Field Club 1880—81,
    Append., t. 4, f. 5.
Lag. orbignyana BRADY, 1884, For. Chall. Rep. Zool. 9, p. 484, t. 59, ff. 1 (18?) 24—25 (26?)
      »   BALKW. & MILLETT, 1884, For. Galway, Journ. micr. Sc. 3, t. 3, f. 1.
BRADY, PARK. & JONES, 1888, For. Abrohlos Bank, Trans. Zool. Soc. London 12, 7, p. 222, t. 44, f. 20.

* forma tri- aut quadrigona:
Lag. trigono-marginata PARK. & JONES, 1865, Phil. Trans. 155, t. 18, f. 1.
WRIGHT, 1877, Rec. For. Down & Antrim; Proc. Belfast, nat. Field Club 1876—77, App. t. 4, f. 8.
L. trigono-orbignyana BALKW. & MILLETT, 1884, For. Galway, Journ. micr. Sc. 3, t. 3, f. 10.
L. quadrigono-orbignyana BALKW. & MILLETT, ibid., t. 4, f. 2.

2. lenticularis, tricarinata, plus minusve regulariter costata aut semilineata; interdum trigona.

L. pulchella BRADY, 1870, Brack. Water For. A. N. H. (4) 6, p. 294, t. 12, f. 1.
BALKW. & WRIGHT, 1885, Dubl. Rec. Foramf., Trans. R. Ir. Ac. Sc. 28, p. 342, t. 12, f. 19 (trigona).
BALKW. & MILLETT, 1884, For. Galw., Journ. micr. Sc. 3, t. 2, ff. 1, 3, ibid., t. 3, f. 11 (trigona).

Hab. ad Sartorö, prope Bukken sinu Norvegiæ metr. 40—70, rara (NORMAN), pusilla, fere circularis, mm. 0.30.

APPENDIX.

3. ovalis aut elliptica, compressa aut subcompressa, interdum carinata, sæpe apiculata.

Entosol. marginata v. quadrata WILLIAMS, Rec. For. Gr. Brit., t. 1, f. 27, a
    REUSS merito ad hanc relata.
Aliquando trigona = Lag. trigono-oblonga BALKW. & MILLETT, Journ. microsc. &
    nat. Sc. 1884, 3, p. 13, t. 3, f. 4.

L. lucida WILL.

Entosolenia marginata v. **lucida** WILL., 1858, Rec. For. Gr. Brit., p. 10, f. 22.
L. lucida REUSS, 1862, Lagenid., Wien. Ak. S. Ber. 46, p. 324, t. 2, f. 25.
Fissurina acuta, oblonga, apiculata Rss, 1862, ibid., pp. 339—340, t. 6, f. 85; t. 7, fl. 89—91.
Lag. lucida BALKW. & MILLETT, 1884, For. Galway; Journ. micr. Sc. 3, t. 2, f. 7.

Specimina sub nomine Lag. bicarinata TERQU. a Cel. Domino Jos. WRIGHT benignissime mihi communicata, valde compressa, collo abbreviato, a Lag. lucida WILL. nisi carina duplicata non diversa.

Hab. ad oras Anglicas.

L. lagenoides WILL.

Tab. XIII, fig. 752.

Compressa, endo-ectosolenica, marginata-alata, alæ tubulis transversis instructæ; interdum nunc seriatim punctata, nunc striata; aliquando trigona.
A Lag. ornata WILL. nisi alis hujus lacunosis distingvenda.

Entosolenia marginata v. **lagenoides** WILL., 1858, Rec. For. Gr. Brit., p. 11, t. 1, fl. 25—26.
Lag. lagenoides Rss, 1862, Lagenid., Wien. Ak. S. Ber. 46, p. 329, t. 2, ff. 27—28.
    »   BRADY, 1884, For. Chall. Rep. Zool. 9, p. 479, t. 60, ff. 6, 7, 9, 12—14.
BALKW. & WRIGHT, 1885, Rec. Dubl. For., Trans. R. Ir. Ac. Sc. 28, p. 341, t. 12, f. 22 (trifacialis).
BALKW. & MILLETT, 1884, For. Galway, Journ. micr. Sc. 3, t. 2, f. 11.

Hab. mare Spetsbergiæ metr. 1,900 rara. Long. 0.28 mm.

L. striatopunctata PARK. & JONES.

Tab. XIII, fig. 753.

Pyriformis, costata, costæ uni-biseraliter foveolatæ, nunc ecto-, nunc endosolenica aut endo-ectosolenica; collum interdum alatum.
Figg. 753, 753 a: e mari Spetsbergico profund. metr. 180.

L. sulcata var. **striatopunctata** PARK. & JONES, 1865, N. Atl. & Arct. Oc., Phil. Trans. 155, p. 350, t. 13, ff. 25—27.
? **L. foveolata** Rss, 1862, Lagenid., Wien. Ak. S. Ber. 46, p. 332, t. 5, f. 65.
L. seriato-granulosa Rss, 1870, For. Sept. Thon. Pietzpuhl; Wien. Ak. S. Ber. 62, p. 468; v. SCHLICHT, tab. 36, f. 20.
L. striatopunctata BRADY, 1884, For. Chall. Rep. Zool. 9, p. 468, t. 58, f. 37 (40 var. torquatæ BRADY magis propinqua).
BALKW. & WRIGHT, 1885, Rec. Dubl. Foramf., Trans. R. Irish Ac. Sc. 28, p. 339, t. 14, f. 19.

Hab. mare Spetsbergicum metr. 180 rara. Long. mm. 0.46—0.50.

GLOBIGERINA D'ORB.

G. bulloides D'ORB.

Tab. XIV, figg. 754—762.

Subtrochoidea aut plano-convexa, anfractibus 2—3, segmentis anfr. ultimi 3—5, plerumque 4, plus minusve inflatis subglobosis; apertura umbilicali nunc vestibulum magnum subquadratum exhibente, nunc valde coarctata. Testa sæpissime areolata, spinosa aut spinoso-squamosa, poris nunc magnis nunc minutis. ˙

Fig. 754: exempla duo minuta; *a, a:* facies orales; *b, b:* facies spirales.

Fig. 755: minima; *a:* facies oralis; *b:* facies spiralis.

Fig. 756: magis trochoidea; *a:* facies oralis; *b:* facies marginalis; omnes e mari Spetsbergensi profund. metr. 180—1,300.

Fig. 757: e mari Atlantico boreali profund. metr. 500; *a:* facies spiralis; *b:* facies marginalis.

Figg. 758—759: magis robusta, e mari Germanico profund. metr. 180; *a, a:* facies orales; *b:* facies marginalis.

Fig. 760: e mari Norvegico profund. metr. 200; poris crebris minutis; *a:* facies oralis.

Fig. 761: in G. dubiam EGGER vergens, e mari Caraibico profund. metr. 540; *a:* facies oralis; *b:* facies marginalis; *c:* exemplar magis trochoide- ejusdem.

Fig. 762: in G. æquilateralem BRADY transiens, apertura valde coarctata; *a:* facies marginalis et oralis; *b:* facies umbilicalis; e mari Azorico profund. metr. 530.

D'ORB., 1826, Tab. méth., An. Sc. nat. 7, p. 277, No. 1, Mod. No. 17, 76.
GOËS, 1882, Ret. Rhizop. Carib. Sea, Sv. Vet. Akad. Handl. 19, 4, p. 92, ubi synonymi; t. 6, f. 203.
BRADY, 1884, Chall. Rep. Zool. 9, p. 333, t. 77, 79, f. 3—7.
BRADY, PARK. & JONES, 1887, Abrohlos Bank, Trans. Zool. Soc. Lond. 12, 7, t. 45, f. 15.

Hab. extra orns Sveciæ occidentales metr. 20—90; mare boreale atlant. & glaciale metr. · 500—1,870 abundans; ad Spetsbergias metr. 350—1,250 (pygmæn).   Diam. mm. 0.30—0.66.

Orbulina magnum globigerinæ segmentum embryonale est habenda maturum, qvod prolem generare possit, quamobrem in Orbulina hanc inclusam haud raro invenis. Embryones istæ aut uniloculares, matri similes, sunt aut multiloculares, globigerinæformes. Orbulinæ singulæ maximæ numquam fere embryones includunt, quasi nimis maturæ, ætate confectæ, seniles.

Naturali de connexu inter Orbulinam et Globigerinam interse præterea dissentierunt auctores. M. SCHULTZE et v. REUSS (Böhm. Gesellsch. Wissensch. Sitz. Ber. 1861, p. 13) Orbulinam segmentum Globerinæ ultimum, disjunctum habuerunt. ALCOCK (Literar. & Philos. Soc. Manch. Mem. (3) 3, 1868, p. 175) Orbulinam sarcodi glomeramen Globigerinæ extra matrem collectum, ubi postea testa circumdatum ratione probabili censet.

Dom. SCHLUMBERGER (Comptes rendus d. Seances de l'Acad. Sciences April 1884) primus sententiam nostræ fere parem proposuit. Si Globigerinæ singulæ Orbulinas, et hæ Globigerinas pariunt, genera distincta non juste habendæ.

Dom. SCHLUMBERGER (l. cit), BRADY (Challeng. Rep., p. 607, et præterea SCHACKO (Wiegmanns Arch. 49, p. 428) orbulinas magnitudine media (mm. 0.70) embryones globigeriniformes procreandi præcipue esse compotes observarunt, quæ potentia pro magnitudine superante decrescat.

formæ affines:

1. crassa, subglobosa, paullum depressa, anfractibus plerumque 2 indistinctis, segmentis anfr. ultimi 4, suturis indistinctis; apertura umbilico-suturali, semilunari, angusta.

G. pachyderma (Ehrenb.) Brady, 1884, Chall. Rep. Zool. 9, p. 600, t. 114, f. 19—20.
Glob. bulloides v. **borealis** Brady, 1881, Österr. Ungar. N. pol. Exped., Wien. Ak. Dkschr. 43,
p. 103; A. M. Nat. Hist. (5) 1, p. 435, t. 21, f. 10.

Hab. in sinubus Groenlandiæ passim; plerumque pygmæa.

2. facie spirali plana aut subplana, facie umbilicali tumida, anfractibus 3—4, segmentis anfr. ultimi 3—4, apertura plerumque suturali marginali, aut semimarginali magna, semilunari, interdum valde angusta fere obliterata, poris minutis; superficie interdum scrobiculata; sæpe Pulleniæ obliqueloculatæ similis.

G. inflata D'Orb., Tab. XIV, figg. 763—765.

Fig. 763: facies umbilicalis; *a:* facies marginalis cum apertura; *b:* facies spiralis; e mari Azorico profund. metr. 530.

Fig. 764: facies spiralis; *a:* facies oralis; *b:* facies marginalis; e mari Germanico profund. metr. 180.

Fig. 765: facies spiralis; *a:* facies umbilicalis; *b:* facies marginalis, cum apertura angustissima; e mari Azorico profund. metr. 80—120.

D'Orb., 1839, For. Canar., p. 134, t. 2, ff. 7—9.
Glob. bulloides var. **inflata** Park. & Jones, 1865, North Atlant. & Arct. Oc., Phil. Trans. 155,
p. 367, t. 16, f. 16—17.
Owen, 1867, Surface Fauna Mid. Ocean, Journ. Lin. Soc. (Zool.) 9, p. 148, t. 5, f. 13.
Brady, 1884, Chall. Rep. Zool. 9, p. 601, t. 79, ff. 8—10.

Hab. mare Germanicum et sinum Codanum metr. 180, sat frequens. Diam. mm. 0.55—0.60.

APPENDIX.

3. trochoidea, spira plus minusve prominens, anfractibus plerumque 3, singulo 3 camerato, cameris globosis, inflatis; apertura umbilicali, circulari quadrilatera aut semilunari, suturali; præterea aperturæ suturales semilunatæ paucæ supra facie spirali sparsæ.
Color dilute ruber, interdum albidus; a Glob. triloba Reuss non sat distincta; spira interdum applanata, interdum pyramidalis.

G. rubra D'Orb., Tab. XIV, fig. 766.

Fig. 766: facies umbilicalis; *a:* facies spiralis; *b:* facies marginalis, exempli spira minus elevata; e mari Caraibico.

D'Orb., 1839, For. Cuba, p. 94, t. 4, f. 12—14.
Glob. rubra Vanden Broeck, Etude forminf. Barbade, Ann. Soc. Belg. microsc. 2, p. 125, t. 3, ff. 9—10.
          »   Brady, 1884, Chall. Rep. Zool. 9, p. 602, t. 79, ff. 11—16.
    »   Brady, Park. & Jones, 1887, Abrohlos Bank, Trans. Zool. Soc. Lond. 12, 7, t. 45, f. 12.
? Glob. canariensis D'Orb., 1839, For. Canar., p. 133, t. 2, ff. 10—12.

Hab. mare Caraibicum locis certis abundans metr. 500 (Goës). Diam. mm. 1.25.

4. subsymmetrice planospiralis anfractibus 1½—3, ultimo sæpe distante, 5—6 camerato, cameris subglobosis, singulis orificio late semilunari umbilicum versus aperto instructis.

G. æquilateralis Brady, Tab. XIV, fig. 767.

Fig. 767: facies spiralis; *a:* facies umbilicalis; *b:* facies marginalis; e mari Azorico profund. metr. 530.

(1879) 1884, Chall. Rep., p. 605, t. 80, ff. 18—21.
Glob. hirsuta d'Orb., 1839, For. Cuuar., p. 133, t. 2, ff. 4—6.
Goës, 1881, Rel. Rhizop. Carib. Sea, Sv. Vet. Akad. Handl. 19, 4, t. 6, ff. 201—202.

Hab. mare Caraibicum metr. 500 (Goës); ad Azores metr. 500 (Smitt & Ljungman). Diam. mm. 0.50—1.

5. magis depressa, anfractu ultimo segmentis 5—7 prædita; præterea ut in typica, a qua non sat distincta, neque a Gl. dubia Egger, magis inflata, distingui debet; poris sæpe minutis.

G. cretacea d'Orb., 1840, For. Craie bl. Paris, Mém. Soc. geol. France 4, p. 34, t. 3, ff. 12—14.
Goës, 1881, Rel. Rhizop. Carib. Sea, Sv. Vet. Akad. Handl. 19, 4, t. 6, ff. 204—206.
Brady, 1884, Chall. Rep., p. 596, t. 82, f. 11 (fig. 10 potius ad typum referenda).

Hab. mare caraibicum metr. 500 passim, pygmæa mm. 0.15—0.30.

6. depressa, inflata, anfractu ultimo 3—4-camerata, segmentis globosis, ultimo (interdum et penultimo) conico-mitræformi, ab anfractu penultimo sæpe disjuncto; apperturæ aliquot magnæ.

G. sacculifera Brady, 1884, Chall. Rep., p. 604, t. 80, ff. 11—17; t. 82, f. 4.
Goës, 1882, Rel. Rhizop. Carib. Sea, Sv. Vet. Akad. Handl. 19, 4, t. 6, ff. 197—200.
Glob. helicina Carpenter, 1862, Introduct. Stud. Foram., t. 12, f. 11.

Hab. mare Caraibicum metr. 300—700. Long. mm. 1.50.

7. globosa, compacta, anfractibus sæpe indistinctis, ultimo cameris 3, rare 4 prædito; apertura umbilicali magna, suturali femilunari, præterea aperturæ minores suturales in facie spirali sparsæ (interdum obsoletæ). Anfractus interni sæpe trigonaliter dispositi.

G. conglobata Brady, Tab. XIV, figg. 768—769.

Fig. 768: facies umbilicalis; *a:* facies spiralis aperturis fere obsoletis; e mari Azorico.

Fig. 769: facies umbilicalis; *a:* facies spiralis, aperturis fere obsoletis aut minutis; e mari Caraibico.

Brady, 1884, Chall. Rep., p. 603, t. 80, ff. 1—5, t. 82, f. 5.
Goës, 1882, Retic. Rhizop. Carib. Sea, Sv. Vet. Akad. Handl. 19, 4, t. 6, f. 196.
Brady, Park. & Jones, 1888, Abrohlos Bank, Trans. Zool. Soc. Lond. 12, 7, t. 45, f. 18.

Hab. mare Caraibicum metr. 180—400 (Goës); extra Azores insulas metr. 500 (Smitt, Ljungman).

SPHÆROIDINA D'ORB.

S. bulloides D'ORB.

Tab. XIV, fig. 770.

Globosa, interdum ovato-pyramidalis, anfractibus apparentibus 2 indistinctis, interdum 3, segmentis anfractus ultimi 3—4, tumidis, magnis; apertura rima semilunari aut semicirculari, suturali sæpe valvulata.

Sæpe nitida, poris minutis.

Fig. 770: facies oralis; a: facies spiralis; e mari Germanico.

D'ORB., 1826, Tab. méth., Ann. Sc. nat. 7, p. 267, No. 1, Mod. 65.
? **Sexloculina Haueri** CZJZEK, 1847, For. Wieuerbeck., Hnidng. Nat. Wiss. Abh. 2, p. 149, t. 13, ff. 35—38.
GOËS, 1882, Ret. Rhiz. Carib. Sen, Sv. Vet. Akad. Handl. 19, 4, p. 89, t. 6, ff. 190—193.
BRADY, 1884, Chall. Rep. Zool. 9, p. 620, t. 84, ff. 1—7.
BRADY, PARK. & JONES, 1888, Abrohlos Bank, Trans. Zool. Soc. Lond. 12, 7, t. 45, ff. 9—11.

Hab. mare Germanicum metr. 180; Österfjord Norvegiæ metr. 180—360 passim (NORMAN). Diam. mm. 0.50—0.60.

PULLENIA PARK. & JONES.

P. sphæroides D'ORB.

Tab. XIV, figg. 771—772.

Subglobosa, paullum compressa, nautiloidea, nitida, segmentis 4, rare 5, suturis tenuibus, vix impressis; apertura rima angusta, transversa marginalis, suturalis.

Figg. 771—772; facies laterales; a, a: facies marginales; e mari Spetsbergico.

Nonion. **sphæroides** D'ORB., 1826, Tab. méth., Ann. Sc. nat. 7, p. 293, No. 1, Mod. 43.
Pull. bulloides RSS, 1870; v. SCHLICHT, Septar. Thon. Pietzpubl, t. 20, ff. 1—2.
» **sphæroides** BRADY, 1884, Chall. Rep., p. 615, t. 84, ff. 12—13.
» BRADY, PARK. & JONES, 1888, Abrohlos Bank, Trans. Zool. Soc. Lond. 12, 7, t. 43, ff. 21, 24.

Hab. in Österfjord Norvegiæ metr. 670, passim (NORMAN), extra Spetsbergiam 72° Lat. bor. metr. 360 (v. OTTER). Diam. mm. 0.35.

forma affinis:

Magis compressa, segmentis apparentibus 4—5, suturis plerumque magis impressis, prætera ut in præcedente, qua sæpe paullo major.

P. quinqueloba RSS, Tab. XIV, fig. 773, c.

v. REUSS, 1851, Septar. Thon. Berlin; Zeitschr. deutsch. geol. Gesellsch. 3, t. 5, f. 31.
P. compressiuscula RSS, v. SCHLICHT, t. 20, ff. 5—6.
GOËS, 1881, Ret. Rhizop. Carib. Sen, Sv. Vet. Akad. Handl. 19, 4, t. 8, ff. 248—249 (segm. extern. 4)
BRADY, 1884, Chall. Rep., p. 617, t. 84, ff. 14—15.
BRADY, PARK. & JONES, 1888, Abrohlos Bank, Trans. Zool. Soc. Lond. 12, 7, t. 23, ff. 22, 23.

Hab. cum precedente. Diam. mm. 0.50.

PLANORBULINA d'Orb.

P. lobatula Walk. & Jacob.

Tab. XV, fig. 774.

Sæpissime affixa, plano- aut concavo-convexa, sæpe sinuose undata; facies aboralis
s. spiralis anfractum solum ultimum plerumque exhibens, interdum obsolete umbilicata,
segmentis 6—8; facies oralis anfractus 2—3 præbens; apertura rima suturalis, marginalis-
extraumbilicalis.

Pori sæpe exigui, sparsi, interdum præcipue faciei oralis obsoleti.

Variat: magis compacta, segmentis anfractus ultimi usque ad 12.

Fig. 774: facies aboralis; *a:* facies marginalis; *b:* facies oralis, typicæ.

Nautilus lobatulus Walk. & Jac., 1798, Adams Essay microscop. (Kanmacher), p. 642, t. 14, f. 36.
Truncatulina tuberculata d'Orb., 1826; Tabl. méth., Ann. Sc. nat. 7, p. 279, No. 1, Mod. 37.
Planulina incerta d'Orb., ibid., p. 280, No. 1.
Truncat. variabilis Reuss, 1870; v. Schlicht, For. Septar. Thon. Pietzpuhl, t. 21, ff. 12—23, 27—29.
   »   **lobatula** Brady, 1884, Chall. Rep. 9, p. 660, t. 92, f. 10; t. 93, ff. 1, 4—5; t. 115, ff. 4—5.
   »   »   Brady, Park., Jones, 1887, Abrohlos Bank, Transact. Zool. Soc. London 12, 7, t. 45, f. 26.
   »   »   Sherborn & Chapm., 1886, London Clay; Journ. R. micr. Soc. (2) 6, t. 16, f. 12.
Synonymiam ceteram vide Goës, Ret. Rhizop. Caribb. Sea, Sv. Vet. Akad. Handl. 19, 4, p. 96.

Obs. Nautilus faretus Ficht. & Moll., 1803, Test. microsc., p. 64, t. 9, ff. g—i est forsan forma distincta,
    poris majoribus.

Anomalina irregularis Terqu., 1886, For. & Ostrac. Islande, Bullet. Soc. Zool. France 1886, p. 333, t. 11,
    ff. 14—15.
? **Truncat. globulosa** Terqu., ibid., t. 11, ff. 12—13.
Truncat. lobatula Fornasini, 1893, For. marne messinesi, Mem. Accad. Sc. Instit. Bologna (5) 3, t. 2,
    ff. 15—16.

Hab. ad oras Sveciæ occidentales et Norvegiæ in algis lapidibusque adnata, profund. metr. 5—50 sat frequens;
    mare Groenlandicum profund. metr. 100—300. Diam. mm. 1.30 et ultra.

formæ affines:

1.   cameris plus minusve irregulariter dispositis.

P. variabilis d'Orb.

Truncatul. variabilis d'Orb., 1826, Tabl. méth., Ann. Sc. nat. 7, p. 279; ?1839, For. Canar. p. 135,
    t. 1, f. 29.
Planorb. truncata Egger, 1857, Mioc. Ortenburg; Leonh. u. Bronns Jhb. 1857, p. 280, t. 10,
    ff. 15—17.
Truncat. tuberosa Park., Jones, Brady, 1871, Nomenclature Foram., A. M. N. Hist. (4) 8, p. 177,
    t. 12, f. 138.
   »   **innormalis** Costa, 1854, Pal. Nap. 2, t. 21, f. 11.
   »   **variabilis** Brady, 1884, Chall. Rep. 9, p. 661, t. 93, ff. 6, 7.
   »   -   Brady, Park., Jones, 1887, Abrohlos Bank, Transact. Zool. Soc. London 12, 7,
    t. 45, f. 17.

Hab. c. præcedente raro præsertim in tropicis.

2. plus minusve elate conica, sæpe subcarinato-limbata, compacta, suturis interdum obscuris:

P. refulgens (Mtf.) d'Orb. Tab. XV, figg. 775—776.

Fig. 775: facies aboralis; *a:* facies marginalis; *b:* facies oralis.
Fig. 776: facies marginalis alius magis typicæ.

Truncatulina refulgens d'Orb., 1826, Tab. méthod. Ann. Sc. Nat. 7, p. 279, t. 13, ff. 8—11; Mod. 77, (cameris quam in nostra forma fere duplo frequentioribus).
Brady, 1884, Chall. Rep. 9, p. 659, t. 92, ff. 7—9.

Hab. in sinubus Norvegiæ, Spetsbergiæ, Groenlandiæ, metr. 120—250 passim. Diam. mm. 1—1.50.

3. deplanata, tennis, suturis sæpe impressis, interdum limbatis, margine sæpe acuto, poris magnis; segmentis arcuatis; inter lobatulam et rotulam d'Orb. medium tenens; a Rotal. Schlönbachi Rss, 1862 (Nordd. Hils. u. Gault, Wien. Ak. S. Ber. 46, p. 84, t. 10, f. 5) vix distincta:

P. Wüllerstorfi Schwag. Tab. XV, fig. 777.

Fig. 777: facies aboralis; *a:* facies marginalis; *b:* facies oralis.

Anomal. Wüllerstorfi Schwag., 1866, Novara Reise, Geol. 2, p. 258, t. 7, ff. 105—107.
Anomal. tenuissima Rss, 1855, Tert. Sch. nördl. u. mittl. Deutschl., Wien. Ak. S. Ber. 18, p. 244, t. 5, f. 60 et
Truncat. costata v. Hken, 1877, For. Clávul. Szab. Sch. p. 73, t. 9, f. 2, non sat distinctæ.
» **Wuellerstorfi** Brady, 1884, Chall. Rep. Zool. 9, p. 662, t. 93, ff. 8, 9.
» **Uhlig**, 1886, Golliz. Carpat. Alttertiär. Mikr. Faun., Jhb. K. K. geol. Reichsanst. 36, p. 174, fig. 3.

Hab. mare glaciale et Atlanticum metr. 1,780—4,630 sat frequens (v. Otter). Diam. mm. 0.60—1.50.

4. planoconvexa aut depresse biconvexa, facies aboralis sæpe plus minusve impressa aut centro subplano, suturis sæpe impressis; centrum faciei oralis interdum subumbonatum; forma nostra marginata; inter typicam et Ungerianam d'Orb. A Planorb. lobatula nisi spira faciei aboralis obsolete aparente et segmentis anfr. ultimi plerumque pluribus distincta:

A Rotal. Kalembergensi, Truncat. Boueana d'Orb., non sat distincta, illa libera, hæc affixa.

P. Akneriana d'Orb. Tab. XV, figg. 778—779.

Fig. 778: facies aboralis; *a:* facies marginalis; *b:* facies oralis exempli e Farsund, Norvegiæ, profund. metr. 35.
Fig. 779: facies aboralis; *a:* facies marginalis; *b:* facies oralis, exempli e sinu Hardanger, Norvegiæ, metr. 50.

Rotal. Akneriana d'Orb., 1846, Bass. tert. Vienne, p. 156, t. 8, ff. 13—15.
Truncatulina Akneriana Brady, 1884, Chall. Rep., p. 663, t. 94, f. 8 (suturis faciei oralis sublimbatis; facies aboralis sparsim tuberculata.)

Hab. in sinubus Bahusiæ, Norvegiæ et Spetsbergensibus metr. 30—350 raro. Diam. mm. 0.75.

P. Ungeriana d'Orb.

Tab. XV, fig. 780.

Biconvexa aut subplano-convexa, altitudine valde varians, nunc tennis, marginata, nunc magis distensa; segmentis anfractus ultimi 9—14; facies aboralis plus minusve — interdum obsolete — umbonata, umbone interdum granoso, prædita; facies oralis nunc obsolete umbilicata, nunc umbonata; poris magnitudine variantibus, formæ tropicæ minoribus, crebrioribus. Forma nostra lobatulæ proxima, a qua interdum vix distinguenda.

Fig. 780: facies aboralis; a: facies marginalis; b: facies oralis exempli formæ emaciatæ.

Rotalina Ungeriana d'Orb., 1846, Bass. tert. Vienne, p. 157, t. 8, ff. 16—18.
    »   **constricta** v. Reuss, 1861, Kreide v. Rügen: Wien. Ak. S. Ber. 44, p. 329, t. 6, f. 7: t. 7, f. 1.
Truncat. Ungeriana Rss, 1870; v. Schlicht, Sept. Thon. Pietzpuhl, t. 21, ff. 1—3.
Planorb. Ungeriana Goës, 1882, Retic. Rhizop. Carib. Sea, Sv. Vet. Akad. Handl. 19, 4, p. 100, t. 7, ff. 234—240 (magis distensa; et variet. affixa); forma distensa a (Rotalia) Truncat. rosea (d'Orb.) Brady, Chall Rep., p. 667, t. 96, f. 1 vix distincta.
Truncat. Ungeriana Brady, 1884, Chall. Rep. 9, p. 664, t. 94, f. 9.
     pygmæa (Hken) Brady ibid., p. 666, t. 95, ff. 9—10 vix specifice distincta.
Planorb. Ungeriana Sherb. & Chapm., 1886, London Clay., Journ. R. microsc. Soc. (2) 6, t. 16, f. 16.

Hab. ad oras Bahusiæ Sveciæ profund. metr. 50 minus frequens, præterea mare Germanicum et sinum Codanum metr. 180—890 (Théel), mare Atlant. boreale metr. 1,780 (Jakmain).

P. coronata Park. & Jones.

Tab. XV, figg. 781—783.

Truncato-conica aut plano-spiralis, discoidea aut plano-convexa, umbilicis concavis, depressis plerumque plicatis aut rugosis, spira anfractuum laterum amborum nuda aut subnuda, rugis tamen obscura, margine late truncato aut subrotundato, segmentis anfr. ultimi 8—9; apertura marginalis, suturalis, pori interdum obstructi, magnitudine medii.

Facies aboralis interdum valde angusta, quare forma Pl. refulgenti valde propinqua exstat, ex qua forsan sit derivata.

Fig. 781: facies oralis, aporata.

Fig. 782: facies marginalis.

Fig. 783: facies aboralis exempli alius refulgenti similis; a: facies marginalis ejusdem.

Anomalina coronata Park. & Jones, 1857, For. Coast Norway, A. M. N. H. (2) 19, p. 294, t. 10, ff. 15, 16.
    »   »   Brady, 1864, Rhizop. Shetland, Trans. Lin. Soc. 24, p. 469, t. 48, f. 13.
Planorb. farcta v. **coronata** Park. & Jones, 1865, North Atl. & Arct. Oc., Phil. Transact. 155, p. 383, t. 14, ff. 7—11.
Anomalina coronata Brady, 1884, Chall. Rep., p. 675, t. 97, ff. 1—2.
    »   »   Fornasini, 1893, For. marne messinesi; Mem. Accad. Sc. Instit. Bologna (5) 3, t. 2, f. 17.

Hab. ad oras Norvegiæ profund. metr. 100—530 haud rara. Diam. 1—1.20.

P. ariminensis d'Orb.

Tab. XV, figg. 784—785.

Plana, compressa, spiralis, segmentis anfract. ultimi 8—11, suturis sæpe limbatis, interdum impressis, margine truncato aut subcarinato, latus utrumque alterum alteri fere simile, apertura marginali, suturali.

Fig. 784: facies aboralis; *a:* facies marginalis; *b:* facies oralis exempli e mari Azorico profund. metr. 530.

Fig. 785: facies oralis; *a:* facies marginalis, e mari Germanico profund. metr. 180.

(Planulina) ariminensis d'Orb., 1826, Tab. méth., Ann. Sc. nat. 7, p. 280, t. 14, fi. 1—3 bis, Mod. 49.
Planorb. tuberosa v. Ariminensis Goës, 1882, Retic. Rhizop. Carib. Sen, Sv. Vet. Akad. Handl. 19, 4, p. 98, t. 7, ff. 228—233.
Anomalina ariminensis Brady, 1884, Chall. Rep. 9, p. 674, t. 93, ff. 10—11.
    »   Brady, Park., Jones, 1888, Abroblos Bank, Trans. Zool. Soc. Lond. 12, 7, t. 45, ff. 20—22.

Hab. mare Germanicum profund. metr. 180 rara; præterea ad Azores insulas metr. 530 (Smitt & Ljungman). Diam. 0.87—1.

P. mediterranensis d'Orb.

Tab. XV, fig. 786.

Applanata, affixa, juvenis regulariter spiralis, adulta cameris irregulariter cyclice dispositis; facies affixa sive oralis sola spiram exhibens; hyalina, cameris centralibus fulvis.

Fig. 786: facies aboralis; *a:* facies marginalis.

d'Orb., 1826, Tab. méth., Ann. Sc. nat. 7, p. 280, Mod. 79; 1846, Bass. tert. Vienne, p. 166, t. 9, ff. 15—17.
Planorb farcta var. vulgaris Goës, 1882, Retic. Rhizop. Carib. Sen, Sv. Vet. Akad. Handl. 19, 4, p. 97, t. 7, f. 227. Synonymia ibid., ab qua Pl. variabilis d'Orb. abstrahenda est.
   » mediterranensis Brady, 1884, Chall. Rep., p. 656, t. 92, ff. 1—3.

Hab. ad oras Sveciæ et Norvegiæ profund. metr. 50—180, haud frequens. Diam. mm. 0.50—0.80.

GYPSINA Carter.

G. inhærens Schultze.

Tab. XV, fig. 787.

Affixa, plus minusve depressa, irregulariter expansa, stadium juvenile segmentis sæpe planospiraliter dispositis, præterea irregulariter acervatis, polygonis aut semiglobosis, apertura nulla, poris magnis, uti apud Planorbulinas, in facie affixa sæpe obsoletis, margine excepto.

Fig. 787: facies superna sive libera; *a:* facies marginalis; *b:* facies affixa, exempli e mari Bahusiæ.

Acervulina inhærens Schultze, 1854, Organ. Polythal. p. 68, t. 6, f. 12.
Tinoporus lucidus Wright, 1877, Rev. For. Down & Antrim; Proc. Belfast. nat. Field. Club 1876—77; App. p. 105, t. 4, ff. 4—5.
Gyps. inhærens Brady, 1884, Chall. Rep. 9, p. 718, t. 102, ff. 1—6.

Hab. ad oras Bahusiæ profund. metr. 40—50 in testaceis affixa, passim. Lat. mm. 1.80.

G. vesicularis PARK. & JONES var. intermedia.

Tab. XV, fig. 788.

Plerumque affixa, truncato-subconica, ovoidea aut subglobularis, foveolato-areolata; cameris magnitudine variantibus, poris nunc obsoletis aut minoribus, nunc majoribus. Inter Gyps. globulum (RSS) BRADY et vesicularem PARK. & JONES medium tenet.

Fig. 788: facies libera; *b:* facies affixa; *c:* sectio transversa.

Orbitolina vesicularis PARK. & JONES, 1866, Nomenclat. Foram., A. M. N. H. (3) 6, p. 31.
Tinoporus lævis CARP., 1861, Research. For., Phil. Transact. 150, p. 559, t. 21, ff. 1—3.
     vesicularis CARP., 1862, Introd. Foramf., p. 224, t. 15, ff. 1—4.
     lævis BRADY, 1864, Rhizop. Shetl., Trans. Lin. Soc. 24, p. 470, t. 48, f. 17.
Gyps. vesicularis BRADY, 1884, Chall. Rep. 9, p. 718, t. 101, ff. 9—12.

Hab. ad oras Bohusiæ et Norvegiæ profund. metr. 20—70 haud rara.   Diam. 1.60.

RUPERTIA WALLICH.

R. stabilis WALLICH.

Tab. XV, fig. 789.

Affixa, producto-spiralis, obovata aut clavata, plus minusve pedunculata, basi sub-expansa; anfractibus 2—3, raptim increscentibus, interdum 5, segmentis anfractuum 4—6 singulorum, sæpe plus minusve inflatis. Apertura, fissura semilunaris, marginalis, obliqua interdum in vestibulum umbilicale magnum spectans. Pori nunc crebri nunc sparsi, nunc minores nunc magni, interdum in series dispositi. Nunc subhyalina, nunc calcarea nitida, nunc valde incrustata suturis obsoletis.

Fig. 789: facies oralis; *a:* facies lateralis; *b:* basis affixa, subtus; *c:* sectio trans-versa anfr. medii.

WALLICH, 1877, On a new sessile Foramf. fr. the N. Atlantic., A. M. N. H. (4) 19, p. 501, t. 20.
SCHLUMBERGER, 1883, Note s. For. nouv. golfe de Gascogne; Feuille des jeunes naturalistes 13, p. 27, t. 2, ff. 6—8.
BRADY, 1884, Chall. Rep., p. 660, t. 98, ff. 1—12.

Hab. mare Spetsbergicum metr. 1,300; Atlanticum boreale profund. metr. 1,750 (LINDAHL); altitudo mm. 1.50.

PATELLINA WILLIAMS.

P. corrugata WILL.

Conica, patelliformis, orbicularis aut subovata, infra concava, segmentis plus minusve indistincte biserialibus, amplectentibus; stadio larvali apicali irregulariter spirali, anfractibus circiter 2; segmentis sequentibus septis incompletis, ut costis aut rugis verticalibus externe indicatis, divisis. Corpus testæ internæ ex cameris irregularibus constitutum.

WILLIAMS, 1858, Rec. For. Gr. Brit., p. 46, t. 3, ff. 86—89.
CARPENTER, 1862, Introduct. Stud. Foramf., p. 230, f. 13, ff. 16—17 (segmentis cyclicis).
BRADY, 1884, Chall. Rep. 9, p. 634, t. 86, ff. 1—7.
PARK. & JONES, 1865, North. Atl. & Arct. Oc., Phil. Trans. 155, t. 15, f. 29.

Hab. ad Groenlandiam profund. metr. 90 rara (NORMAN).

DISCORBINA PARK. & JONES.

D. Berthelotiana D'ORB.

Tab. XV, fig. 790.

Depressa complanata tenuis, hyalina, facie aborali leniter concava nunc umbilicata, nunc paullum impressa, anfractu ultimo præcedentem nunc totum nunc ex parte obtegente; segmentis anfr. ultimi 5—9; facie orali excavata, interdum subumbilicata; suturis, hujus interdum uti costis prominentibus, distinctis; poris minutis.

Fig. 790: facies aboralis; *a:* facies marginalis; *b:* facies oralis exempli e mari Germanico.

Rosalina D'ORB., 1839, For. Canar., p. 135, t. 1, ff. 28—30.
GOËS, 1882, Retic. Rhizop. Carib. Sea, Sv. Vet. Akad. Handl. 19, 4, p. 105, t. 8, ff. 266—268.
BRADY, 1884, Chall. Rep., p. 650, t. 89, ff. 10—12.
Discorb. baconica v. HKEN, 1875, For. Czab. Szab. Sch., Mittheil. Jhrb. ungar. geol. Anst. 4. p. 76, t. 10, f. 3 (major, limbata, carinata).
　　　rarescens BRADY, 1884, Chall. Rep. 9, p. 651, t. 90, ff. 2—3 (carinata vel carinato-alata).
　　Berthelotl BRADY, PARK. & JONES, 1887, Abrolhos Bank, Trans. Zool. Soc. Lond. 12, 7, t. 46, ff. 7—8.

Hab. mare Germanicum profund. metr. 180 passim; ad oras Norvegicas metr. 90—540 (NORMAN, v. YHLEN). Diam. mm. 0.50—0.60.

D. parisiensis D'ORB.

Tab. XV, fig. 791.

Convexa aut planoconvexa, tenuis, obtuse marginata, spira nunc aperta nunc anfractu ultimo celata, segmentis hujus 6—12, arcuatis, plus minusve angustatis; facie infera s. orali tuberculis minutissimis, in radios dispositis, instructa; suturis sæpe sinuosis.

Forma nostra suturis paullum impressis, interdum valde deplanata, fere discoidalis, poris sparsis, magnitudine mediis.

Variat segmentis angustioribus magis numerosis = D. opercularis d'ORB.

Fig. 791: facies aboralis; *a:* facies marginalis; *b:* facies oralis.

Rosalina D'ORB., 1826, Tab. method; An. Sc. nat. 7, p. 271, No. 5, Mod. 39.
PARKER, JONES, BRADY, 1866, Crag. For., Palæogr. Soc. 19, t. 2, ff. 13—15.
WRIGHT, 1877, For. Down & Antrim, Proc. Belfast Nat. Field Club 1876—77, Append., p. 105, t. 4, f. 1.
BRADY, 1884, Chall. Rep. 9, p. 648, t. 90, ff. 5, 6, 9—12.

Hab. ad insulam Hven profund. metr. 20, rara (THÉEL & TRYBOM). Diam. mm. 0.45.

D. rosacea D'ORB.

Tab. XV, fig. 792.

Trochoidea, late conica, hyalina, obtuse marginata, subtus subplana aut paullum ex-cavata, anfractibus 3—4, ultimo 4—7 segmentato, facie orali saepe lobulis umbilicalibus instructa ("asterisante"), poris minutis aut mediis saepe marginalibus; a Discorbina (Rotalia) turbone D'ORB. nisi statura magis depressa non distincta, quod valde varians.

Variat segmentis angustioribus, semilunaribus, amplectentibus, suturis saepe limbatis.

Fig. 792: facies aboralis; a: facies marginalis; b: facies oralis; c mari Bahusiae.

Rotalia D'ORB., 1826, Tab. méth., Ann. Sc. nat. 7, p. 273, No. 15, Mod. 39.
Asterigerina planorbis D'ORB., 1846, For. Russ. tert. Vienne, p. 205, t. 11, ff. 1—3 (segmentis angustis amplectentibus).
Rotal. mamilla WILLIAMS, 1858, Rec. For. Gr. Brit., p. 54, t. 4, ff. 109 —111 (segment. amplectentibus, poris marginalibus limbata).
? Discorb. stellata Rss, 1857, Steinsalzablag. Wieliczka, Wien. Ak. S. Ber. 55, p. 101, t. 5, f. 1 (paullum limbata).
? Discorb. squamula Rss, ibid., t. 5, f. 2 (subtus limbatis suturis).
Discorb. rosacea BRADY, 1884, Chall. Rep. 9, p. 644, t. 87, ff. 1, 4 (cufr. Discorb. turbonis figuram ibidem tab. 87, f. 8).
Obs. Discorb. orbicularis (TERQUEM) BRADY, Chall. Rep.. p. 647, t. 88, ff. 4—8, segmentis valde angustis amplectentibus. in unfractu ultimo 3—4, ab Asterigerina planorbi EGGER 1857, Mioc. Ortenburg, Leonh. & Bronns Jhb. 1857, p. 281, t. 11, ff. 8—10 nec a Rosal. Auberii D'ORB., 1839, For. Cuba, p. 94, t. 4, ff. 5—8 sat distincta, est varietas D. rosaceae habenda. Inter ambas medium tenet D. rosacea GOËS, 1882, Retic. Rhizop. Carib. Sea, Sv. Vet. Akad. Handl. 19, 4, p. 105, t. 8, ff. 251—257.
Rosalina squamiformis Rss, 1854, Kreide Ostalp., Wien. Ak. DkSchr. 7, p. 69, t. 26, f. 2 (ab qua nostra nisi lobulis umbilicalibus destituta, non distincta).
Rotalina nitida WILLIAMS, 1858, Rec. For. Gr. Brit., p. 54, t. 4, ff. 106—108 verosimiliter ad D. rosaceam referenda; a BRADY ad Rotaliam relata: Chall. Rep., p. 705; Synopsis Brit. Rec. For., Journ. Roy. microsc. Soc. 1887, p. 293.

Hab. in Skagerack sinus Codani profund. metr. 50 rara (H. THEEL). Diam. mm. 0.50—0.60.

D. globularis D'ORB.

Tab. XV, fig. 793.

Trochoidea, facie aborali convexa, spiram totam exhibente, anfractibus 2—3—3½, segmentis ultimi 5—7, margine obtuso, poris majoribus; facie orali saepe aporata, plerum-que concava aut subplana, interdum irregulariter impressa et tumida, umbilico irregulari; suturis saepe impressis, interdum sinuosis.

Color testae plerumque flavidus.

Fig. 793: facies aboralis; a: facies marginalis; b: facies oralis.

Rosalina D'ORB., 1826, Tab. méth., An. Sc. nat. 7, p. 271, t. 13, ff. 1—4, Mod. 69.
Rot. semiporata EGGER, 1857, LEONH. & BRONNS Jhb. 1857, p. 276, t. 8, ff. 1—3.
» concamerata WILL. (ex parte), 1858, Rec. For. Gr. Brit., p. 53, t. 4, ff. 104—105.
Disc. vesicularis v. globularis PARK. & JONES, 1865, North Atl. & Arct. Oc., Phil. Transact. 155, p. 386, t. 14, ff. 22—23.
» globularis BRADY, 1884, Chall. Rep. 9, p. 643, t. 86, ff. 8, 13.

Hab. ad oras occidentales Sveciae et Norvegiae inter algas profund. metr. 3—50, vulgaris. Diam. mm. 1—2.

Variat statura minore, poris exiguis:

Discorbina (Rosalina) Vilardeboana, araucana D'ORB., Tab. XVI, fig. 796; exemplum e mari Azorico; *a:* facies marginalis; *b:* facies oralis.

D'ORB., 1839, Voy. Amér. mérid. 5, p. 44, t. 6, ff. 13—18.
BRADY, Chall. Rep., p. 645, t. 86, ff. 9—12; t. 88, f. 2.

D. obtusa D'ORB.

Tab. XV, figg. 794—795.

Planoconvexa, tumidiuscula, **anfractibus faciei aboralis** $1^1/_2$—$2^1/_2$, ultimo segmentis 5—8, spiram nunc subtegente, nunc spiram totam nudam revelante, segmentis tumidiusculis, margine obtuso, aut rotundato; facie orali subplana tuberculis umbilicalibus et striis radiantibus prædita, interdum aporata; poris mediis aut minutis; facies aboralis interdum tuberculata.

Fig. 794: facies aboralis; *a:* facies marginalis; *b:* facies oralis.

Fig. 795: juvenis; *a:* facies marginalis; *b:* facies oralis.

Discorbinæ pulvinatæ BRADY, Chall. Rep. Zool. 9, t. 88, f. 10, facie aborali tuberculata, valde propinqua, forsan eadem.

Rosalina obtusa D'ORB., 1846, Boss. tert. Vienne, p. 179, t. 11, ff. 4—6.
Discorbina vesicularis var. **obtusa** PARK. & JONES, 1865, N. Atlant. & Arct. Oc., Phil. Trans. 155, p. 386, t. 14, ff. 18—19.
? **Discorb. obtusa** BRADY, 1884, Chall. Rep. Zool. 9, p. 644, t. 91, f. 9 (sine striis tuberculisque).

Hab. ad Spetsbergiæ metr. 40 passim in Grantia. Diam. 0.50—1.80.

PULVINULINA PARK. & JONES.

P. repanda FICHT. & MOLL.

Tab. XVI, fig. 801.

Rataliformis, biconvexa aut subplano-convexa, facies spiralis anfractus 3 exhibens, segm. aufr. ultimi 6—10, suturis sæpe sublimbatis; facies oralis nunc vestibulo magno aperturali instructa, nunc apertura coarctata umbone obtecta, suturis sæpe "asterisantibus".

Faciei ambæ interdum scabræ sicuti apud varietatem concameratam MONTAG.

Fig. 801: facies aboralis; *a:* facies marginalis; *b:* facies oralis, exemplum e mari Azorico.

Nautilus repandus FICHT. & MOLL., 1803, Testac. microsc., p. 35, t. 3, ff. a—d.
Pulv. repanda GOËS, 1882, Ret. Rhizop. Carib. Sea, Sv. Vet. Akad. Handl. 19, 4, p. 110, t. 8, ff. 276—282.
BRADY, 1884, Chall. Rep., p. 684, t. 104, f. 18.
> **exigua** BRADY, ibid., p. 696, t. 103, ff. 13—14, non sat distincta.
? **P. repanda** SHERB. & CHAPM., 1886, London Clay, Journ. R. Micr. Soc. (2) 6, t. 16, f. 18.

Hab. ad Azores insulas profund. metr. 50—90 (SMITT & LJUNGMAN). Diam. mm. 2.25.

forma affinis:

Valde limbata, sæpe plus minusve scabra aut tuberculata:

P.(Rotalina)**concamerata** (MONTAG) WILLIAMS, 1858, Rec. For. Gr. Brit., p. 52, t. 4, ff. 102—103.
Pulv. repanda var. **concamerata** BRADY, 1884, Chall. Rep. 9, p. 685, t. 104, f. 19.

Hab. mare Anglicum.

P. punctulata D'ORB.

Tab. XVI, figg. 797—800.

Affixa, expansa, plano- aut concavo-convexa, depressa, segmentis aufr. ultimi 6—8, irregulariter inflatis; facies oralis suturis sinuatis, dentatis aut laceratis, obsolete tuberculata, interdum poris magnis instructa.

Inter congeneres maxima, juvenes suturis sæpe sublimbatis. Interdum pygmæa, hyalina nitida.

Fig. 797: facies aboralis; *a*: facies marginalis; *b*: facies oralis, exempli e mari Germanico.

Fig. 798: juvenis e mari Azorico profund. metr. 600—780.

Fig. 799: e mari Germanico.

Fig. 800: pygmæa e mari Azorico profund. metr. 530.

Rotalia D'ORB., 1826, Tab. méth., Ann. Sc. nat. 7, p. 273, No. 25, Mod. 12.
Rosalina vesicularis PARK. & JONES, 1857, For. Coast of Norway, A. M. Nat. H. (2) 19, p. 292, t. 10, ff. 22—24.
Pulv. repanda v. punctulata PARK. & JONES, 1865, N. Atl. & Arct. Ocean, Philos. Transact. 155, p. 394,
            t. 14, ff. 12, 13.
  »   repanda PARK. & JONES, BRADY, 1866, Crag. For.; Palæogr. Soc. 19, t. 2, ff. 22—24.
  »   punctulata BRADY, 1884, Chall. Rep. 9, p. 685, t. 104, f. 17.

Obs. P. sacculata PARK. & JONES, 1876 (A. M. Nat. H. (4) 17, p. 284, ff. 1—3) ab hac non sat distincta
      videtur.

Hab. in mari Germanico in Oculinis affixa, metr. 350 nec non in sinubus Norvegicis profund. metr. 90 (NORMAN).
      Diam. 3—4 mm.

formæ affines:

1. affixa planoconvexa, suturis præsertim faciei aboralis late limbatis, facies oralis
    umbilico sæpe lobato.

A juvenibus P. punctulata nisi limbatione lata suturarum non distincta; suturis juvenum tenuiter limbatis:

P. concentrica PARK. & JONES, Tab. XVI, figg. 802—803.

Fig. 802: facies aboralis; *a*: facies marginalis; *b*: facies oralis exempli e sinu Bergensium Norvegiæ.

Fig. 803: e mari Azorico profund. metr. 75.

Pulv. concentrica BRADY, 1864, Rhizop. Shetland, Trans. Lin. Soc. 24, p. 470, t. 48, f. 14; 1884,
            Chall. Rep., p. 686, t. 105, f. 1.
  »   UHLIG, 1886, Westgaliz. Karpath. Alttertiär., Jhb. K. K. geol. Reichsanst. 36, p. 190,
            t. 3, ff. 3, 4.

Hab. in sinubus Norvegiæ passim profund. metr. 90—180 (LILLIEBORG, NORMAN).

P. dispansa Brady.

Tab. XVI, figg. 804—806.

Nunc affixa nunc libera, plus minusve irregulariter lobata, segmentis plus minusve irregulariter dispositis, spira obtecta verrucosa; facies oralis plerumque lævis poris magnis sparsis prædita.

Juvenes magis regulariter constructæ.

Fig. 804: magis regularis; *a*: facies marginalis; *b*: facies oralis.

Fig. 805: juvenis.

Fig. 806: valde irregularis; *a*: facies marginalis; *b*: facies oralis.

Pulv. **dispansa** Brady, 1884, Chall. Rep. 9, p. 687, t. 115, f. 3.

Hab. ad Azores insulas metr. 50—600 (Smitt, Ljungman). Diam. 2.25.

P. Karsteni Rss.

Tab. XVI, fig. 807.

Trochoidea, anfractibus $3\frac{1}{2}$—4, segmentis anfr. ultimi 7—8, semiamplectentibus; facies oralis convexa aut subplana, suturis rectis impressis, rare limbatis, umbone obsoleto; margine sublobato, limbato.

A Pulv. Schreibersii d'Orb. non sat distincta.

Fig. 807: facies aboralis; *a*: facies marginalis; *b*: facies oralis.

Rotal. **Karsteni** Rss, 1855, Kreide Mecklenb., Zeitschr. deut. geol. Gesellsch. 7, p. 273, t. 9, f. 6.
Pulvin. **Karsteni** Brady, 1864, Rhizop. Shetl., Transact. Lin. Soc. 24, p. 470, t. 48, f. 15.
»   **repanda** var. **Karsteni** Park. & Jones, 1865, North. Atlant. & Arct. Oc., Philos. Transact. 155, p.
            396, t. 14, ff. 14, 15, 17; t. 16, ff. 38—40.
»   **karsteni** Brady, 1884, Chall. Rep. Zool. 9, p. 698, t. 105, ff. 8, 9.

Hab. in sinubus Spetsbergiæ metr. 180 passim, ad Groenlandiam metr. 20, mare Atlant. boreale metr. 270 (Lindahl). Diam. mm. 0.60—0.80.

P. elegans d'Orb.

Tab. XVI, fig. 808.

Trochoidea, plus minusve æqualiter biconvexa, marginata, anfractibus 3—4, segmentis anfract. ultimi 7—9, latere utroque sæpe umbonato, suturis plane aut elevate limbatis; facies oralis sæpe »stellata», apertura sæpe rima apicalis marginis segmenti ultimi.

Variat suturis non limbatis. Recens testa vitrea, hyalina, albide variegata.

Fig. 808: facies aboralis; *a*: facies marginalis; *b*: facies oralis.

Rotal. **elegans** d'Orb., 1826, Tab. méth., Ann. Sc. nat. 7, p. 276, No. 54.
Pulv. **elegans** Goës, 1882, Ret. Rhiz. Carib. Sea; Sv. Vet. Akad. Handl. 19, 4, p. 111, t. 8, ff. 283--285
Rotalina **pleurostomata** Schlumberg. For. d. Golfe Gascogne, Feuille d. Jeunes natur. 13, p. 27, t. 3, f. 5.

Pulv. Partschiana Rss, 1870; v. Schlicht, t. 20, ff. 23—25, 29—31.
` **partschiana, elegans** Brady, 1884, Chall. Rep. 9, p. 699, t. 105, ff. 3—6.
> Brady, Park., Jones, 1887, Abrohlos Bank, Trans. Zool. Soc. Lond. 12, 7, t. 46, f. 2.
> Fornasini, 1893, For. marne messin; Mem. Accnd. Sc. Instit. Bologna (5) 3, t. 2, f. 18.

Hab. mare Atlanticum boreale metr. 1,750 rara (Lindahl). Diam. mm. 0.90; ad Spetsbergiam, Groenlandiam pygmæa.

P. auricula Ficht. & Moll.

Tab. XVI, figg. 809—810.

Ovalis plus minusve elongata, inæqualiter biconvexa, compressa, sæpe marginata, anfractibus 1¹/₂—2, segmentis anfr. ultimi 7—10, segmento ultimo facie orali subventricosa, umbilicum plerumque obtegente; color sæpe flavescens.

Var. segmento ultimo decurrenti spiram superante = Valvul. cordiformis Costa; ? Rotal. deformis d'Orb. etc.

Fig. 809: facies aboralis; a: facies marginalis; b: facies oralis.
Fig. 810: forma Brongniartii d'Orb.; b: facies oralis.

Nautilus auricula var. α et β Ficht. & Moll., 1803, Test. microsc., p. 108, t. 20, ff. a—f.
Pulv. auricula Goës, 1882, Ret. Rhizop. Carib. Sea; Sv. Vet. Akad. Handl. 19, 4, p. 109, t. 8, ff. 273—275, ubi synonymia completa; Obs. Rotal. contraria Rss, 1851 (Sept. Thon. Berlin., Ztschr. deut. geol. Gesellsch. 3, p. 76, t. 5, f. 37 est forsan Bulimina quædam; Rot. cristellarioides Rss, For. Crag. Anvers., Bullet. Acad. Belg. (2) 15, p. 154, t. 3, f. 44, est forsan Non. turgida Williams.
> **oblonga** Brady, 1884, Chall. Rep. 9, p. 688, t. 106, ff. 4—5.

Hab. mare Germanicum metr. 180; ad oras Norvegiæ metr. 130 (Lilljeborg, Norman). Long. mm. 1.

APPENDIX.

P. Schreibersii d'Orb.

Trochoidea, subhemisphærica, anfractibus 4—6, spira umbone hemisphærico plerumque subobtecta; segmentis anfr. ultimi 6—8 subamplectentibus; facies oralis convexa aut subplana, suturis plerumque stellato-limbatis.

Rotalina Schreibersii d'Orb., 1846, Bass. tert. Vienne, p. 154, t. 8, ff. 4—6.
> **badensis** Czjzek, 1847, Haid. Naturw. Abh. 2, p. 144, t. 13, ff. 1—3.
Pulv. elegans v. **trochus** Goës, 1882, Ret. Rhizop. Carib. Sea, Sv. Vet. Akad. Handl. 19, 4, p. 112, t. 8, ff. 286—288.
> **Schreibersii** Brady, 1884, Chall. Rep. 9, p. 697, t. 115, f. 1.
>   > Brady, Park. & Jones, 1888, Abrohlos Bank, Trans. Zool. Soc. Lond. 12, 7, t. 46, f. 4.

Hab. mare Carnibicum sat frequens metr. 500—600 (Goës).

ROTALINA (Lmck.) d'Orb.

R. Beccarii Lin.

Tab. XVI, fig. 811.

Helicoidea, biconvexa aut subplano-convexa, margine rotundato, anfractibus 3—4, anfr. ultimo segmentis 8—12; facie aborali plerumque magis elevata, suturis sæpe limbatis nunc impressis, nunc superficiem testæ æquantibus; facie orali magis depressa, area centrali plus minusve rugosa, verrucosa, suturis irregulariter excavatis, rimosis, quasi incrustatis; umbilico valde granuloso, profundo, interdum subumbonato. Apertura, rima marginalis, suturalis.

Formæ emaciatæ occurrunt, rugis verrucisque faciei oralis valde obsoletis.

Fig. 811: facies aboralis; *a:* facies marginalis; *b:* facies oralis exempli magis elevati e mari Bahusiæ.

Nautilus Beccarii testæ apertura obovata, anfractibus contiguis torulosis, geniculis insculptis, Lin. Syst. Nat.
 Ed. X, 1758, p. 710.
» » Montag. 1808, Test. Brit., Supplem. p. 74, t. 18, f. 4.
Rotalia Beccarii d'Orb. (Turbinulina), 1826, Tabl. méth., Ann. Sc. nat. 7, p. 275, No. 42, Mod. 74 (deplanata).
 » **tortuosa, corallinarum** d'Orb., 1826, ibid., No. 40, 48, Mod. 75.
Rosal. Parkinsoniana d'Orb., 1839, For. Cuba, p. 99, t. 4, ff. 25—27.
 » **Catesbyana** d'Orb., ibid., p. 99, t. 4, ff. 22—24.
 » **viennensis** d'Orb., 1846, Foss. tert. Vienne, p. 177, t. 10, ff. 22—24.
 » » Egger, 1857, Mioc. Ortenburg, Leonh. & Bronn Jhb. 1857, p. 277, t. 8, ff. 11—13.
 » **Amaliæ, radiata** Costa, 1854, Pal. Napol. 2, p. 254, 255, t. 21, ff. 12—13.
 » **inflata** Seg., 1862, Rhiz. Catania, Accad. Gioenia Atti (2) 18, p. 22 (separ.), t. 1, f. 6.
Rotal. punctato-granulosa Seg., 1879, For. terz. Prov. di Reggio; Atti R. Accad. Lincei (3) 6, p. 147,
 t. 13, f. 37.
 Beccarii Williams, 1858, Rot. For. Gr. Brit., p. 48, t. 4, ff. 90—92.
Rosal. Mackeyi Karr., 1864, Grünsand, N. Zeeland, Novar. Exped. geol. Th. 1, 2, p. 82, t. 16, f. 14.
Rotal. Beccarii Park. & Jones, 1865, N. Atl. & Arct. Oc., Phil. Trans. 155, p. 388, t. 16, ff. 29—30.
 » » Park. & Jones, Brady, 1866, Crag. For., Palæogr. Soc. 19, t. 2, ff. 19—21.
 » » Uhlig, 1883, For. rjäsan'shen Ornatenthone, K. K. geol. Reichsanst. Jhb. 33, p. 773, t. 8, f. 8.
 » » Brady, 1884, Chall. Rep. 9, p. 704, t. 107, ff. 2—3 (ubi Synonymia vetusta reperitur).
 » » Wright, Chalk. For. Keady Hill; Proc. Belfast Nat. Field Club 1884—85, Append. t. 27, f. 15.

Hab. ad oras Bahusiæ metr. 2—150 frequens, diam. mm. 1.10.

R. Soldanii d'Orb.

Tab. XVI, fig. 812.

Planoconvexa plus minusve tumida, nitida, interdum submarginata, facie aborali planiuscula, spira interdum paullum elevata, sæpe subumbonata, anfractibus 3—4, segmentis anfr. ultimi 8—11; facie orali sæpe valde turgida elevata, umbilicata, plerumque lævi. A Rot. orbiculari d'Orb. nisi facie orali magis tumida non diversa.

Fig. 812: facies aboralis; *a:* facies marginalis; *b:* facies oralis exempli e mari Atlant. boreali profund. metr. 1,750. Fig. supra *a:* faciem aboralem exempli alius præbens.

Rotalia (Gyroidina) Soldanii D'ORB., 1826, Tab. méth., An. Sc. nat. 7, p. 278, No. 5, Mod. 36; 1846, Buss. tert. Vienne, p. 155, t. 8, ff. 10—12.

  »   **umbilicata** D'ORB., 1839, Craie blanche d. Paris, Mém. Soc. geol. Fr. 4, p. 32, t. 3, ff. 4—6 (non sat distincta).

  »   **turgida** v. HAGENOW, 1842, Rügens Kreide Verstein; Leonh. & Br. Jhb. 1842, p. 570, t. 9, f. 22.

  »   **conoidea** CZJZ., 1847, For. Wien. Beck., Haid. naturwissen. Abh. 2, p. 145, t. 13, ff. 4—6.

  »   **Girardana** Rss, Sept. Thon. Berl., Zeitschr. deut. geol. Gesellsch. 3, p. 73, t. 5, f. 34; v. SCHLICHT, t. 20, ff. 11—13.

  »   **Beccarii** v. **Soldanii** PARK. & JONES, N. Atl. & Arct. Oc., Phil. Trans. 155, p. 389, t. 16, ff. 31—33.

  »   **nitidula** SCHWAG, 1866, For. Kar. Nikobar; Novara Reise geol. Th. 2, p. 263, t. 7, f. 110.

  »   **Soldanii** v. HKEN, 1875, For. Clàv. Szab. Sch., p. 80, t. 9, f. 7.

  »   »   BRADY, 1884, Chall. Rep. 9, p. 706, t. 107, ff. 6—7 (conf. Rotal. orbicularis (D'ORB.), ibid., t. 115, f. 6, quæ a figura, t. 107, ff. 6—7, non distingvenda est).

Hab. in Stoksund Norvegiæ metr. 140—180 (NORMAN); mare Atlanticum boreale metr. 1,750 (LINDAHL) haud frequens.

POLYSTOMELLA D'ORB.

P. arctica PARK. & JONES.

Tab. XVI, fig. 813.

Nautiloidea, disciformis, subumbonata, latere utroque subplano, margine rotundato, segmentis 8—14. Suturæ sæpe limbatæ impressæ, ostiis canaliculorum interseptalium poriformibus utrinque (interdum uniserialibus) instructæ, pontibus septalibus destitutæ; area umbilicalis ostiis sparsis provisa.

Apertura marginalis, suturalis porosa aut rima margine superno crenulata, ostiis interdum nonnullis supramarginalibus superpositis.

Fig. 813: exemplum e mari Spetsbergensi profund. metr. 150; *b:* facies marginalis.

Polystomella arctica PARK. & JONES, M. S. 1864.

BRADY, 1864, Rhiz. Shetl., Trans. Lin. Soc. 24, p. 471, t. 48, f. 18; 1878, Ret. Rhizop., N. Pol. Exped. 1875 —76, A. M. N. H. (5) 1, p. 437, t. 21, f. 13.

P. crispa v. **arctica** PARK. & JONES, 1865, N. Atl. & Arct. Oc., Phil. Trans. 155, p. 401, t. 14, ff. 25—30.

BRADY, 1884, Chall. Rep. 9, p. 735, t. 110, ff. 2—5.

BALKW. & WRIGHT, 1885, Dublin For., Trans. R. Irish Acad. Sc. 28, p. 353, t. 13, f. 27.

Hab. mare Spetsbergicum & Groenlandicum metr. 140—180 sat frequens; ad Newfoundland et Nova Semjla. Diam. mm. 1.50.

P. sibirica GOËS.

Tab. XVII, fig. 814.

Lenticularis, biconvexa plus minusve applanata, interdum subcarinata, cameris circiter 24, angustatis, umbonibus interdum circumscriptis, ostiis septalibus plerumque in binas series interdum in unam seriem dispositis; apertura poris minutis exhibita.

Fig. 814: facies lateralis; *a:* facies marginalis.

Hab. ad Nova Semjla, Maçotschkin Sharr metr. 30 sat frequens (STUXBERG, THÉEL 1875). Diam. mm. 3—4.

P. striato-punctata (FICHT. & MOLL.) PARK. & JONES.

Tab. XVII, figg. 815—816.

Discoidea, lateribus subplanis, umbilicis subdepressis, interdum absentibus, umbonibus obsoletis substitutis; segmentis 8—12 subinflatis, suturis paullum impressis, pontibus septalibus brevissimis, foveolis plus minusve distinctis, interdum fere evanescentibus; apertura marginalis, suturalis, porosa aut crenulata, poris septalibus interdum additis paucis.

Testa interdum obsolete squamoso-areolata, sæpe hyalina, poris magnitudine variantibus; (testa aliquando subarenacea?).

Fig. 815: hyalina; a: facies marginalis; b: in luce transeunte visa; e Gullmaren sinu Bahusiæ.

Fig. 815 c—f: duæ majores e mari Groenlandico profund. metr. 18; c: umbone obsolete granuloso.

Fig. 815 g—i: e mari Spetsbergico metr. 180—550, ostiis septalibus provisa.

Fig. 815 k—l: e mari N. Semjlu.

Fig. 815 m—r: pygmææ e mari Bahusiæ profund. metr. 80; m: superficiem areolatam præbens.

Fig. 815 s—u: e mari Spetsbergico.

Fig. 816: forma subarenacea? Haplophragmio similis; ex eodem loco; a: facies marginalis.

? **Nautilus striatopunctatus** FICHT. & MOLL., 1803, Test. micr., p. 61, t. 9, ff. a—c (striata, inflata sine pontibus septalibus).
Polyst. Poëyana D'ORB., 1839, For. Cuba, p. 55, t. 6, ff. 25—26.
Geoponus stella borealis EHRENB., 1839, Jetzt lebende Thierarten der Kreidebild., Abh. K. Ak. Wiss. Berlin, p. 132, t. 1, ff. a—g.
Polyst. Haurina D'ORB., 1846, Bass. tert. Vienne, p. 122, t. 6, ff. 1—2.
? **Polyst. rugosa, obtusa** D'ORB., ibid., t. 6, ff. 3—6 (segment. 15—20).
Polyst. Antonina, Lietori D'ORB., ibid., p. 128. t. 6, ff. 17—22.
  » **stella borealis, gibba, venusta** SCHULTZE, 1854, Organ. Polythal., p. 60, 67, ff. 1—9.
Non. heteropora, densepunctata, Polyst. subcarinata EGGER, Mioc. Ortenburg, Leonh. & Bronn Jhb. 1857, p. 299—301, t. 14, ff. 19—25.
Polyst. umbilicatula WILLIAMS, 1858, Rec. For. Gr. Brit., p. 42, t. 3, ff. 81—82
  »   var. incerta WILL., ibid., t. 3, f. 82 a.
Non. excavata SEG., 1862, Rhizop. Catania, Atti Accad. Gioena (2) 18, t. 1, f. 4.
Polyst. latidorsata RSS, 1863, For. Oberburg, Denkschr. 23, p. 10, t. 1, f. 16.
  »   **minuta, discrepans** RSS, 1864, Oberoligocän, Wien. Ak. S. Ber. 50, p. 478, t. 4, ff. 6—7.
  »   **striatopunctata** JONES, PARK. & BRADY, 1866, For. Crag., Pal. Soc. 19, t. 2, ff. 38—39.
  »   **crispa** var. **striatopunctata** PARK. & JONES, 1865, N. Atl. & Arct. Oc. Phil. Trans. 155, p. 402, t. 14, ff. 31—34, t. 17, f. 50.
  »   **minima** SEG., 1879, For. terz. Prov. Reggio, Atti Accad. Lincei (3) 6, p. 333, t. 17, f. 38.
  »   **crispa** var. **Poëyana** GOËS, 1882, Ret. Rhizop. Carib. Sea, Sv. Vet. Akad. Handl. 19, 4, p. 116, t. 6, ff. 301—302.
  »   **striatopunctata** BRADY, 1884, Chall. Rep. 9, p. 733, t. 109, ff. 22—23.
  »   »   BRADY, PARK., JONES, 1887, Abrohlos Bank, Trans. Zool. Soc. Lond. 12, 7, t. 43, f. 17.

Hab. ad oras Sveciæ, Norvegiæ, Groenlandiæ, Spetsbergiæ frequens, metr. 2—1,700. Diam. mm. 0.30—1.10.

forma affinis:

Magis inflata aut carinata, sæpe umbonata.

P. subnodosa v. Münst. (sec. Reuss), Tab. XVII, figg. 817—819.

Fig. 817: e mari Nova Semjla; *a:* facies marginalis oralis.

Fig. 818: pygmæa in typicam vergens; e mari Bahusiæ Sveciæ.

Fig. 819: crassa, e mari Newfoundlandico; *a:* facies marginalis oralis.

Robulina subnodosa v. Münst., Leonh. & Bronns Jhb. 1838, p. 391, t. 3, f. 61.
Nonionina splendida Boll., 1846, Geogr. deut. Ostseeländ., p. 177, t. 2, f. 15.
Polyst. subnodosa Reuss, 1855, Tert. Sch. nördl. u. mittl. Deutschl., Wien. Ak. Sitz. Ber. 18, p. 240,
　　　t. 4, f. 51.
　　》　**exoleta** Costa, 1854, Pal. Napol. 2, t. 19, f. 10.
　　》　**cryptostoma, angulata, Ortenburgensis** Egger, 1857, Mioc. Ortenburg; Leonh. & Bronns
　　　Jhb. 1857, pp. 301, 302, t. 9, ff. 19—20; t. 15, ff. 5　9.
? **Naut. ambiguus** Ficht. & Moll., 1803, Test. micr., p. 62, t. 9, ff. d—f.
? **Polyst. subumbilicalis** Czjz., 1847, For. Wien. Beck., Haid. naturwiss. Abh. 2, p. 143, t. 12,
　　ff. 32—33.
? **Polyst. Ungeri, flexuosa** Rss, 1849, Neue For. Österr., Wien. Ak. Dkschr. 1, p. 369, t. 48, ff. 2—8.

Hab. ad New Foundland rara (Lindahl); Nova Semjla (Stuxberg).

P. crispa Lin.

Tab. XVII, figg. 820—821; 822.

Lenticularis, plus minusve umbonata (aliquando umbonibus destituta), margine attenuato sæpe carinato; segmenta anfractus ultimi 12—24; suturis elevatis, extra marginem ut spinis plerumque prominentibus, pontibus septalibus transversis interse connectis aut subconnectis; apertura marginalis, suturalis, poris constituta.

Fig. 820: exemplum e mari Bahusiæ profund. metr. 80; *a:* facies marg. oralis.

Fig. 821: exemplum e Zostera litorum Bahusiæ; *a:* facies marg. oralis.

Fig. 822: inter P. crispam & striatopunctatam, e sinu Quænang Norvegiæ profund. metr. 200.

Nautilus crispus, testæ apertura, semicordata, anfractibus contiguis, geniculis crenatis, Lin. Syst. nat., Ed. X,
　　　1758, p. 709.
　　》　　》　? Ficht. & Moll., 1803, Test. micr., p. 40, t. 4, ff. d—f, t. 5, ff. a—b (pontibus septal.
　　　omissis); Montagu, 1803, Test. Brit., p. 187, t. 18, f. 5.
Polyst. crispa d'Orb., 1826, Tab. méth., An. Sc. nat. 7, p. 283, Mod. 45.
　　》　**oweniana** d'Orb., 1839, Voy. amér. mérid., p. 30, t. 3, ff. 3, 4.
　　》　**Lanieri** d'Orb., 1839, For. Cuba, p. 54, t. 7, ff. 12—13.
　　》　**Berthelotiana** d'Orb., 1839, For. Canar., p. 129, t. 2, ff. 31—32.
　　》　**ornata, crispa,** ? **flexuosa** d'Orb., 1846, For. Rass. tert. Vienne, p. 125, t. 6, ff. 9—14, p. 127, t. 6,
　　　ff. 15—16 (P. flexuosa in Polyst. striatopunct. v. subnodosa m Münst.
　　　vergens).
　　》　**crispa, decipiens, striolata, ornata** Costa, 1854, Pal. Napoli 2, t. 14, f. 11, t. 19, ff. 13, 15—16.
　　》　**crispa** Park. & Jones, 1865, N. Atl. & Arct. Oc., Phil. Trans. 155, p. 399, t. 14, f. 24; t. 17, f. 61.
　　》　》　Williams, 1858, Rec. For. Gr. Brit., p. 40, t. 3, ff. 78—80 (margine spinosa).
　　》　》　Carpenter, 1862, Introd. Foramf., p. 278, t. 16, f. 4 (margine spinoso).
　　》　》　Brady, Chall. Rep. 9, p. 736, t. 110, ff. 6—7.

Obs. Polyst. regina, Josephinæ, aculeata d'Orb., Rass. tert. Vienne, p. 129—131, t. 6, ff. 23—28, margine aculeis ornato, a typo difficillime distingvendæ.

Hab. ad oras Sveciæ occidentales, metr. 3—80, passim. Diam. mm. 0.80 (plerumque minuta).

NONIONINA d'Orb.

N. umbilicatula Montag.

Tab. XVII, figg. 823—824.

Nitida, interdum pellucida, nunc subdepressa, nunc paullum inflata, umbilicata aut subumbilicata, margine rotundato, segmentis 9—12, suturis vix impressis; facie septali convexa, semilunari; apertura suturali, marginali, angusta, poris mediocribus; interdum foveolis septalibus minutis provisa. Aliquando e particulis arenaceis et calcareis immixtis constructa, Haplophragmio similis.

Fig. 823: e sinu Bergensium Norvegiæ, profund. metr. 200; *a:* facies marginalis oralis.

Fig. 824: calcareo-arenacea e Gullmaren sinu Bahusiæ profund. metr. 70.

Naut. umbilicatulus Montag, 1803, Test. Brit., p. 191, Supplem. 1808, p. 78, t. 18, f. 1.
Non. Soldanii d'Orb., 1846, Bass. tert. Vienne, p. 109, t. 5, ff. 15—16 (in N. pompilioidem vergens).
   »   »   Costa, 1854, Pal. Nap. 2, p. 201, t. 17, f. 11 (idem).
   -     **falx** Czjz, 1847, For. Wien. Beck., Haid. Nat. Wiss. Abh. 2, p. 142, t. 12, ff. 30—31.
   »   **polystoma** Costa, 1854, Pal. Nap. 2, p. 206, t. 14, f. 10.
   »   **crassula** Park. & Jones, 1857, For. Norw. A. M. N. H. (2) 19, p. 286, t. 11, ff. 5—6.
   »   **Barleana** Williams, 1858, Rec. For. Gr. Brit., p. 32, t. 4, ff. 68—69.
   »   **bathyomphala** Rss, 1862, Nordd. Hils u. Gault, Wien. Ak. S. Ber. 46, p. 95, t. 13, f. 1.
Polystom. crispa var. **umbilicatula** Park. & Jones, 1865, N. Atl. & Arct. Oc., Phil. Trans. 155, p. 405, t. 14, f. 42, t. 17, ff. 58—59.
Non. formosa Seo., 1879, Foram. terz. Prov. Reggia, Atti Accad. Lincei (3) 6, p. 63, t. 7, f. 6.
   »   **umbilicatula** Brady, 1884, Chall. Rep. 9, p. 726, t. 109, ff. 8—9.
   »   »   Brady, Park., Jones, 1887, Abrohlos Bank, Trans. Zool. Soc. Lond. 12, 7, t. 43, f. 19.

Hab. ad oras occidentales Sveciæ et Norvegiæ metr. 40—180 frequens, ad Spetsbergiam metr. 180—1,260 freqventissima.

N. depressula Walk. & Jacob.

Tab. XVII, figg. 825—826.

Discoidea, margine plerumque rotundato, aliquando submarginata, umbilicis haud multum aut vix excavatis subgranulosis; interdum subumbonata; segmentis 7—11—14, suturis sæpe impressis, umbilicum versus sæpe stellate — excavatis, aut sublimbatis; plerumque hyalina, poris minutis creberrimis; a Polystomella striatopunctata interdum difficile distingvenda.

Fig. 825: e mari Germanico, in formam sequentem vergens; *a:* facies later. oralis.

Fig. 826: e mari Groenlandico; *a:* facies later. oralis.

Naut. depressulus Walk. & Jac., 1798, Adams Essays microsc., p. 641, t. 14, f. 33.
Non. perforata, granosa, punctata d'Orb., 1846, Bass. tert. Vienne, p. 110, 111, t. 5, ff. 17—22.
   »   **crassula** Williams, 1858, Rec. For. Gr. Brit., t. 3, ff. 70—71.
   »   **affinis** Rss, 1851, Sept. Thon. Berl., Zeitschr. deut. geol. Ges. 3, p. 72, t. 5, f. 32.
Polystom. crispa var. **(Non.) depressula** Park. & Jones, 1865, North Atl. & Arct. Oc., Phil. Trans. 155, p. 403, t. 14, f. 39.

Goës, 1882, Ret. Rhizop. Carlb. Sea, Sv. Vet. Akad. Handl. 19, 4, t. 8, ff. 269—271.
Non. depressula BRADY, 1884, Chall. Rep. 9, p. 725, t. 109, ff. 6—7.

Obs. Nautilus incrassatus FICHT. & MOLL., 1803, Test. min., p. 38, t. 4, ff. a—c, a N. depressula non
    sat distinctus.

Hab. mare Germanicum metr. 180, mare Arcticum metr. 100—1,870, passim. Diam. mm. 0.40—0.45.

forma affinis:

Suturis stellatæ limbatis, umbilico plerumque depresso, poris sæpe quam in typica
    paullo majoribus.

N. stelligera D'ORB., Tab. XVII, figg. 827—828.

Fig. 827: e mari Germanico; stella umbilicali impressa, non sat typica.

Fig. 828: exemplum e Bukken sinu Norvegico.

D'ORB., 1839, For. Canar., p. 128, t. 3, ff. 1—2.
BRADY, 1864, Rhizop. Shetl., Trans. Lin. Soc. 24, p. 471, t. 48, f. 19.
Polyst. crispa v. stelligera PARK. & JONES, 1865, N. Atl. & Arct. Oe., Philos. Trans. 155, p. 404,
    t. 14, f. 40.
Nonionina stelligera BRADY, 1884, Chall. Rep. 9, p. 728, t. 109, ff. 3—5.

Hab. ad Spetsbergiam; in Bukken sinu Norvegico metr. 260 —350 (NORMAN) minuta; mm. 0.30—0.40.

N. Boueana D'ORB.

Tab. XVII, fig. 829.

Late ovalis aut elliptica, compressa, margine subacuto, umbilico vix excavato, interdum tenue granulato; segmentis anfractus ultimi 12—16; suturis sæpe impressis paullum
excavatis aut obsolete stelligeris. A sequente non sat limitanda.

Fig. 829: exemplum e mari Atlantico profund. metr. 1,240; a: facies marg. oralis.

D'ORB., 1846, Bass. tert. Vienne, p. 108, t. 5, ff. 11—12.
BRADY, 1884, Chall. Rep. 9, p. 729; t. 109, ff. 12—13.

Obs. Non. Boueana Rss, 1863, For. Crag d'Anvers, Bullet. Ac. Belg. (2) 15, p. 156, t. 3, ff. 47—48 inter
    hanc et sequentem medium tenet.

Hab. mare Atlanticum lat. 48° boreali longitud. occident. 10° metr. 1,240 passim (LINDAHL). Diam. mm. 0.65.

N. scapha FICHT. & MOLL.

Tab. XVII, fig. 830.

Elongate aut late elliptica, compressa aut inflata, margine obtuse curicato; segmentis
9—12, rare 15, ultimis raptim increscentibus, elongatis aut latis, tumidis; plerumque vix
umbilicata, suturis interdum crenatis; (testa arena aliquando inmixta?)

Figg. 830—830 c: exempla e sinu Gullmaren Bahusiæ.

? Fig. 831: tenue arenacea, Haplophragm. cassidi aliquantum similis, e mari
Groenlandico profund. metr. 40.

Naut. scapha Ficht. & Moll., 1803, Test. micr., p. 105, t. 19, ff. d—f (in N. Boueanam d'Orb. vergens).
Non. Grateloupi, Sloanei, Brownii d'Orb., 1839, For. Cuba, p. 46, t. 6, ff. 6—7, 18—18 bis; t. 7, ff. 22—23.
   communis d'Orb., 1846, Bass. tert. Vienne, p. 106, t. 5, ff. 7—8.
       Park. & Jones, 1857, For. Coast Norway, A. M. N. H. (2), 19, p. 287, t. 11, ff. 7—8.
  » **labradorica** Park. & Jones, Brady, 1866, Crag. For., Pal. Soc. 19, t. 2, ff. 44—45.
Pullenia communis v. Htkn., 1875, For. Cläv. Száb. Sch., p. 59, t. 10, f. 10.
? Non. Frassana Gümb., 1862, Streitberg Schwammlager, Jahresh. Würtemb. naturwiss. Vereins 1862, p. 233, t. 4, f. 5.
Polystom. crispa var. **Scapha** Park. & Jones, 1865 (ex parte(, N. Atl. & Arct. Oc., Phil. Transact. 155, p. 404, t. 14, f. 38.
  » **crassula** v. **Scapha** Goës, 1881, Ret. Rhiz. Carib. Sea, Sv. Vet. Akad. Handl. 19, 4, p. 114, t. 8, ff. 299—300.
Non. Scapha Brady, 1884, Chall. Rep. 9, p. 730, t. 109, ff. 14—15.
  » **labradorica** Daws., 1860, Canad. Natur., p. 192, f. 4; 1871, Gulf & River St. Lawrence A. M. N. H. (4) 7, p. 86, f. 5 (brevis inflata).

Hab. ad oras Sveciæ et Norvegiæ metr. 30—180 frequens; in mari Arctico profund. metr. 180—2,200 vulgaris. Diam. mm. 0.90.

forma affinis:

segmentis ultimis valde elongatis, ultimo sæpe inæquilaterali, spiram fere obtegente:
N. turgida Williams, Tab. XVII, fig. 832.
Fig. 832: exemplum e sinu Codano profund. metr. 230.

Rotalina turgida Will., 1858, Rec. For. Gr. Brit., p. 50, t. 4, ff. 95—97.
? Rotalia cristellarioides Rss, 1863, Crag. d'Anvers, Bull. Ac. Belg. (2) 15, p. 154, t. 3, f. 44.
Polystom. crispa v. (Non.) **turgida** Park. & Jones, 1865, N. Atl. & Arct. Oc., Phil. Trans. 155, p. 405, t. 17, f. 57.
Non. turgida Brady, 1884, Chall. Rep. 9, p. 731, t. 109, ff. 17—19.

Hab. sinum Codanum profund. metr. 100—230 rara; Bukken sinum Norvegicum metr. 260—350 (Norman). Diam. mm. 0.30—0.40.

OPERCULINA d'Orb.

O. ammonoides Gronov.

Tab. XVII, fig. 833.

Planospiralis, complanata, sæpe obtuse- aut truncato-marginata, area umbilicali leniter impressa, anfractibus internis nunc exhibitis, nunc subobtectis; anfractibus 3, segmentis anfr. ultimi numero 9—13, sæpe paullum inflatis, angulo interno sæpe calloso, emarginato aut inciso; suturis plerumque impressis, aut impresso-limbatis.
Fig. 833: anfractibus internis non obtectis; a: facies marg. oralis, exempli e Gullmaren Bahusiæ metr. 75.
Fig. 833 b, a: anfractibus internis umbilico subobtectis; ex eodem loco.
Fig. 833 c: exempl. e Farsund Norvegiæ metr. 75.

Naut. ammonoides Gronov, 1781, Zoophylac. Gronov., p. 282, t. 19, ff. 5—6.
  » **balthicus** Schroet., 1783, Vollst. Einleitung, Gesch. Steine u. Verstelner: 1, p. 20, t. 1, f. 2.
Operc. complanata Park. & Jones, 1857, For. Coast. Norw., A. M. N. H. (2) 19, p. 285, t. 11, ff. 3—4.

Non. elegans WILL., 1858, Rec. For. Gr. Brit., p. 35, t. 3, ff. 74—75.
Nummulina perforata var. (Opero.) **ammonoides** PARK. & JONES, 1865,. N. Atl. & Arct. Oc., Phil. Trans.
          155, p. 398, t. 14, f. 44; t. 17, ff. 62—63.
Operculina ammonoides BRADY, 1884, Chall. Rep. Zool. 9, p. 745, t. 112, ff. 1, 2.

Hab. ad oras occidentales Sveciæ et Norvegiæ profund. metr. 70—260 sat frequens.   Diam. mm. 0.50—0.95.

CORNUSPIRA M. SCHULTZE.

C. foliacea PHIL.

Tab. XVIII, fig. 834.

Planospiralis, valde applanata, anfractibus internis pernumerosis, tenuissimis, sequentibus plus minusve raptim latitudine increscentibus, sequentibus præeuntes haud multum aut vix amplectentibus, lineis plicisque transversis, arcuatis instructis; parte tubi plus matura sæpe longitudinaliter striata aut rugosa; apertura obovata aut rima longa, tubo toto patente; margine interno sæpe ex parte canaliculato aut fisso.

Figg. 834—834 *c:* exempla e mari Groenlandico; *a, c:* facies marg. oralis.
Fig. 834 *d—e:* lentius increscens, e mari Spetsbergensi; *e:* facies marg. oralis.

Orbis foliaceus PHIL., 1844, Moll. Sicil. 2, p. 147, t. 24, f. 26.
Operculina striata, plicata CZJZ, 1847, For. Wien. Beck., Haid. Nat. Wiss. Abh. 2, p. 146, t. 13, ff. 10—13.
Cornuspira planorbis SCHULTZE, 1854, Organ. Polythal., p. 40, t. 2, f. 21.
Operculina ammonitiformis COSTA, 1854, Pal. Nap. 2, p. 209, t. 17, f. 16.
Spirillina foliacea WILL., 1858, Rec. For. Gr. Brit., p. 91, t. 7, ff. 199—201.
Cornuspira foliacea REUSS, 1865, Deutsch. Septarienth., Wien. Ak. DkSchr. 25, p. 121, t. 1, ff. 8—9.
   »   »   PARK. & JONES, 1865, N. Atl. & Arct. Oc., Phil. Trans. 155, p. 408, t. 15, f. 33.
   »   »   JONES, PARK., BRADY, 1866, For. Crag., Pal. Soc. 19, p. 2, t. 3, ff. 50—51.
   »   »   GOËS, 1882, Ret. Rhizop. Carib. Sea, Sv. Vet. Akad. Handl. 19, 4, p. 120, t. 9, f. 308.
   »   »   BRADY, 1884, Chall. Rep. 9, p. 199, t. 11, ff. 5—9.
   »   »   REUSS, 1870; v. SCHLICHT, Sept. Thon. Pietzpuhl, t. 35, ff. 11—12.

Hab. ad oras Scandinaviæ, Groenlandiæ, Spetsbergiæ profund. metr. 10—530, haud rara.   Diam. mm. 3—4.

formæ affines:

1. margine anguste carinato.

C. carinata COSTA.

Operculina carinata COSTA, 1854, Pal. Nap. 2, p. 209, t. 17, f. 15.
C. Bornemanni Rss, 1863, Sept. Thon. Offenbach, Wien. Ak. S. Ber. 48, p. 39, t. 1, f. 3.
  » **carinata** BRADY, 1884, Chall. Rep. 9, p. 201, t. 11, f. 4.

Hab. in sinubus Norvegiæ passim (SARS, NORMAN).

2. minus compressa, anfractibus fere teretibus, quoque suum præcedentem ex parte amplectente.

C. involvens, Rss.

Operculina involvens v. Rss, 1849, Neue Foramf. Österreichs., Wien. Ak. DkSchr. 1, p. 370, t. 46, f. 20.
Cornuspira involvens Rss, Sept. Thon. Offenbach, Wien. Ak. S. Ber. 48, p. 39, t. 1, f. 2.

Cornuspira involvens PARK. & JONES, BRADY, 1866, Crag. For., Pal. Soc. 19, p. 3, t. 3, ff. 52—54.
» » polygyra v. HKEN, 1875, Clavul. Szaboi Sch., Mittheil. n. d. Jhb. K. Ungar.
geol. Anst. 4, p. 19, t. 1, f. 11, t. 2, f. 1.
» » BRADY, 1884, Chall. Rep. 9, p. 200, t. 11, ff. 1—3.

Hab. mari Arctico et Norvegico passim, profund. metr. 30—180; pygmæa.

C. striolata BRADY.

Tab. XVIII, fig. 835.

Anfractibus paucis raptim latitudine increscentibus, quare forma flabelliformis adultarum, plus minusve plicato-striata aut rugoso-squamosa; striis verticalibus sæpe obsoletis; juvenes a var. involvente Rss sæpe vix discernendæ; megasphærica forsan typicæ; inter congeneres maxima.

Fig. 835: magnitudine naturali delineata; e mari Sibirico.

Cornuspira striolata BRADY, 1884, Chall. Rep. 9, p. 202, t. 113, ff. 18—19.

Hab. mare Sibiricum profund. metr. 150 (Dom. THÉEL, STUXBERG). Diam. mm. 35.

SPIROLOCULINA D'ORB.

S. planulata LAMCK.

Tab. XVIII, figg. 836—836 c.

Elliptica, utrinque excavata, margine truncato aut rotundato, cameris planis limbo destitutis; apertura subquadrata aut ovalis lingula simplici aut furcata plerumque instructa.

Fig. 836: angusta, typica, e sinu Bergensium Norvegiæ, 20 metr.
Fig. 836 b: in limbatam transiens; e Gullmaren Bahusiæ metr. 40.

Miliolites planulata LAMCK., 1805, Ann. Mus. 5, p. 352.
Spir. depressa, perforata, ? rotundata D'ORB., 1826, Ann. Sc. nat. 7, p. 298—299, No. 1, 2, 14, Mod. 92.
badenensis, dilatata D'ORB., 1846, Bass. tert. Vienne. pp. 270, 271, t. 16, ff. 13—18.
» depressa var. rotundata WILL., 1858, Rec. For. Gr. Brit., p. 82, t. 7, f. 178.
Freyeri REUSS, 1863, Oberburg. Steiermk, Wien. Ak. Sitz. Ber. 23, p. 7, t. 1, f. 9.
» planulata BRADY, PARK. & JONES, 1866, For. Crag., Pal. Soc. 19, p. 15, t. 3, ff. 37—38.
Mil. planulata PARK. & JONES, 1865, N. Atl. & Arct. Oc., Phil. Trans. 155, p. 408, t. 17, f. 82.
Spir. planulata BRADY, 1884, Chall. Rep. 9, p. 148, t. 9, f. 11.
» » BRADY, PARK. & JONES, 1887, Abrohlos Bank, Trans. Zool. Soc. Lond. 12, 7, p. 214, t. 40, ff. 14—15.

forma affinis:

margine limbato, suturis prominentibus, a typica vix limitanda; nec a Spirol. excavata D'ORB. Bass tert. Vienne, p. 271, t. 16, ff. 19—21 distincta:

S. limbata D'ORB., Tab. XVIII, fig. 837.

Fig. 837: e mari Azorico profund. metr. 70.

D'ORB., 1826, Ann. Sc. nat. 7, p. 299, No. 12.
Spir. limbata BORNEM., 1855, Sept. Thon Hermsd., Zeitschr. deut. geol. Gesellsch. 7, p. 348, t. 19, f. 1.
   >   depressa WILLIAMS, 1858, Rec. For. Gr. Brit., p. 82, t. 7, f. 177.
   >   limbata Rss, 1863, Sept. Thon. Kreuznach, Wien. Ak. S. Ber. 48, p. 64, t. 8, f. 89.
Mil. (Spirol.) limbata PARK. & JONES, 1865, Phil. Trans. 155, p. 409, t. 17, f. 83.
Spir. limbata BRADY, 1884, Chall. Rep. 9, p. 150, t. 9, ff. 15—17.

Hab. ad oras Scandinaviæ profund. metr. 10—70 passim.

MILIOLINA (LAMCK.) WILLIAMS.

M. seminulum LIN.

Tab. XVIII, figg. 838—838 n; Tab. XIX, figg. 840—843.

Oblonga aut subrotundata, compressa aut plus minusve inflata, plerumque nitida, sectione transversa trigona aut triloba; apertura ovata aut subcirculari aut reniformi, lingua bifurcata aut simplici instructa. Segmenta apparentia 5—4, rare 3.

Fig. 838 a, b: e mari Groenlandico profund. metr. 296; b: facies oralis; c, d: sectiones transversæ, dispositionem segmentorum "Sigmoilinæ" præbentes.

Fig. 838 e, f, g: magis triangularis; e mari Azorico profund. metr. 1,780.

Fig. 838 h, i, k: in variet. Auberianam vergens, ex Atlantico boreali profund. metr. 150.

Fig. 838 l, m, n: magis orbicularis, tumida, e mari Bahusiæ profund. metr. 35.

Tab. XIX, fig. 240: e Koster fretis insularum Bahusiæ profund. metr. 2.

Fig. 841: valde inflata fere IV loculina, ex Atlantico boreali, profund. metr. 330.

Fig. 842: valde obliquata, e Skagerack, profund. metr. 700—850.

Fig. 843: in subrotundam transiens; e mari Groenlandico profund. metr. 110.

Serpula seminulum testa regulari ovali, libera, glabra LIN., 1758, Syst. Nat. Ed. X, 1, p. 786.
Synonymiam ceteram vide apud GOËS, 1882, Rel. Rhizop. Carib. Sea, Sv. Vet. Akad. Handl. 19, 4, p. 124 b) et p. 187 d) partim, et BRADY, 1884, Chall. Rep. 9, p. 157, t. 5, f. 6.
Quinqueloc. seminulum (D'ORB.) SCHLUMB., 1893, Mil. golfe Marseille, Mém. Soc. Zool. France 6, p. 208, t. 4, ff. 80—81.

Hab. ad litora Scandinaviæ occidentales et Arctica profund. metr. 2—800 frequens; long. mm. 2.50; nec rfon mare Balticum (pygmæa) passim.

formæ affines:

1. triloculina plus minusve tumida; a formis larvalibus Mil. (Biloc.) simplicis difficile distincta.

M. inornata D'ORB., Tab. XVIII, fig. 839.

Triloculina inornata D'ORB., 1846, Bass. tert. Vienne, p. 279, t. 17, ff. 16—18.
? Triloc. pyrula KARR., 1867, For. Österreichs, Wien. Ak. S. Ber. 55, p. 359, t. 2, f. 7.
Triloc. decipiens REUSS, 1849, Nene For., Wien. Ak. DkSchr. 1, 1, p. 382, t. 49, f. 8.

Hab. extra Väderöarne insulas Bahusiæ profund. metr. 50.

2. segmentis acutangulis aut carinatis, a typo non sat distincta; sæpe lata suborbicularis:

M. Auberiana D'ORB., Tab. XIX, figg. 844—844 e.

Fig. 844 a, b: minus carinata, e mari extra Väderöarne insulas Bahusiæ profund. metr. 45.

Fig. 844 c, d, e: magis carinata, medio impresso, ex eodem loco.

Quinqueloc. **Auberiana** D'ORB., 1839, For. Cuba, p. 193, t. 12, ff. 1—3.
    »   **Gualteriana, Lamarckiana, Cuveriana** D'ORB., ibid., p. 189, 190, t. 11, ff. 14—15, 19—21.
    »   **Buchiana, Ungeriana, Partschii, peregrina** D'ORB., Bass. tert. Vienne, p. 277, 289,
            291, 292, t. 18, ff. 10—12, 22—24, t. 19, ff. 1—6.
? Triloc. **anceps** Rss, 1849, Neue For. Österr., Wien. Ak. DkSchr. 1, p. 383, t. 49, f. 11 (complanata).
Quinqueloc. **inæqualis** D'ORB., 1839, For. Canarcis, p. 142, t. 3, ff. 28—30.
Mil. **seminulum** var. GOES, 1882, Rel. Rhizop. Carib. Sea, Sv. Vet. Akad. Handl. 19, 4, t. 9, ff. 346—347.
    »   **auberiana, cuveriana** BRADY, 1884, Chall. Rep. 9, p. 162, t. 5, ff. 8—9, 12.

Hab. ad oras occidentales profund. metr. 50 rara. Long. mm. 0.50.

3. segmentis angulatis, oblonga:

M. Candeiana D'ORB., Tab. XIX, fig. 845.

Fig. 845: exemplum non sat acute-carinatum e mari Atlantico boreali.

Quinqueloc. **Candeiana** D'ORB., 1839, For. Cuba, p. 199, t. 12, ff. 24—26.
Quinqueloc. **Candeiana,** Kooni Rss, 1855, Kreidegeb. Mecklenburg, Zeitschr. deut. geol. Ges. 7, p. 289,
        t. 11, ff. 6—7.
        longirostra D'ORB., 1846, Bass. tert. Vienne, p. 291, t. 18, ff. 25—27.
?   »   **venusta** KARR., 1868, Mioc. Kostej, Wien. Ak. Sitz. Ber. 58, p. 147, t. 2, f. 6.
?   »   BRADY, 1884, Chall. Rep. 9, p. 162, t. 5, ff. 5, 7.
? Triloc. **carinata** PHIL., 1843, Tert. Verstein. n. w. Deutschl., t. 1, f. 36.

Hab. mare Atlant. Int. bor. 51° profund. metr. 400, rara (LINDAHL).

4. disciformis suborbicularis aut transverse subovata, subplana, margine rotundato; apertura sæpe magna semilunata aut triangularis lingua minuta aut obsoleta; segmentis sæpe irregulariter dispositis, ultimis ambobus margine orali interse sæpe distantibus. Triloculina suborbicularis D'ORB., 1839, For. Cuba, p. 177, t. 10, ff. 9—11 huc forsan referenda, forma striolata.

M. subrotunda WALK. & BOYS., Tab. XIX, figg. 846 - 847 a—h.

Fig. 846: inter seminulum et subrotundam, e Hållö Bahusiæ, profund. metr. 50.

Fig. 847 a—h: formæ variæ e mari Bahusiæ et mari Atlantico boreali e profundis parvis.

Serpula **subrotunda** WALK. & BOYS, 1784, Test. min., p. 2, t. 1, f. 4.
Vermic. **subrotundum** MONTAG, 1803, Test. Brit. 2, p. 521.
Quinqueloc. **subrotunda** D'ORB., 1826, Tab. méth., An. Sc. nat. 7, p. 302, No. 36.
    »   **dilatata** D'ORB., 1839, For. Cuba, p. 192, t. 11, ff. 28—30.
    »   SCHLUMB., 1893, Miliol. Golfe de Marseille, Mém. Soc. Zool. France 6, p. 217,
        t. 3, ff. 70—74, t. 4, ff. 89—90.
    »   **meridionalis** D'ORB., 1839, Voy. Amér. mérid. 5, p. 75, t. 4, ff. 1—3, 10—13.
Mil. **subrotunda** PARK. & JONES, 1865, N. Atl. und Arct. Oc., Phil. Trans. 155, p. 411, t. 15, f. 38.
Triloc. **truncata** KARR., 1864, Leytaknlk Wien. Beck., Wien. Ak. Sitz. Ber. 50, p. 704, t. 1, f. 2.
    »   **dilatata** KARR., 1868, Mioc. Kostej., Wien. Ak. S. Ber. 58, p. 139, t. 2, f. 1.

Miliol. subrotunda BRADY, 1884, Chall. Rep. 9, p. 168, t. 5, ff. 10—11.
?Quinqueloc. implexa TERQU., 1886, For. et Ostrac. d'Islande, Bull. Soc. Zool. France, 1886, p. 335, t. 11, ff. 24—26.
Mll. subrotunda WRIGHT, 1885, For. Belfast, Proc. Belf. nat. Field Club, 1885—86, Append. t. 29, f. 6.

Hab. ad oras occidentales Scandinaviæ profund. metr. 0.5—260; mare Arcticum profund. metr. 20—280 ubique sat frequens. Diam. mm. 0.50—1.25.

5. testa arena incrustata colore brunneo-hepatico, plerumque magis complanata, segmentis terctiusculis; suturis sæpe fere obsoletis, margine rotundato aut subtruncato aut obtuse angulato. Speciem formamque quamque suam formam agglutinantem habere est expectandum.

M. agglutinans PARK. & JONES, Tab. XIX, figg. 848—848 *l*; Tab. XX, fig. 849.

Figg. 848—848 *c:* ex arena minus grandi constructæ, e mari Groenlandico profund. metr. 130—440.

Fig. 848 *f—h:* pauperata, e mari Baltico; an = fusca BRADY.

Figg. 848 *i—l:* e mari Spetsbergico profund. metr. 180; in contortam vergens.

Tab. XX, fig. 849: e mari Norvegico profund. metr. 90—180.

Mil. (Quinqueloc.) agglutinans PARK. & JONES, 1865, N. All. & Arct. Oc., Phil. Trans. 155, p. 410, t. 15, f. 37.
Obs. Quinqueloc. agglutinans D'ORB., 1839, For. Cuba, p. 168, t. 12, ff. 11—13, apertura dentata a nostra distincta, quare nomen formæ nostra in "arenaceam" melius sit transmutandum. M. fuscæ BRADY, 1870, For. tidal. Riv., Ann. M. N. H. (4) 6, p. 286, t. 11, f. 2, exempla a Dom. Cel. Jos. WRIGHT benigne communicata a nostra non specifice distincta.
Quinqueloc. foeda REUSS, 1849, Neue For. Österr., Wien. Ak. DkSchr. 1, p. 384, t. 50, ff. 5—6.
? Miliolina seminulum GOËS, 1884, Ret. Rhizop. Carib. Sca, Sv. Vet. Akad. Handl. 19, 4, t. 9, ff. 319—320.
   »   **agglutinans** BRADY, 1884, Chall. Rep. 9, p. 180, t. 8, ff. 6—7 (magis rudis).
   »   »   BALKW. & WRIGHT, 1885, Rec. Dubl. For., Trans. Roy. Irish. Acad. Sc. 28, p. 325, t. 13, ff. 1—3 (magis rudis).
   »   MOEBIUS, 1880, For. Mauritius, p. 77, t. 3, ff. 4—8.
   »   BRADY, PARK. & JONES, 1888, Abroblos Bank, Trans. Zool. Lond. 12, 7, t. 40, ff. 34—35.

Hab. mare Arcticum metr. 100—350 passim; ad oras Sveciæ et Norvegiæ metr. 90—180 minus frequens; mare Balticum (pygmœa) passim. Long. mm. 1.30.

6. plus minusve elongata, plerumque tri-quadriloculina, a Miliolina ringente var. elongata D'ORB. interdum difficulter distinguenda.

M. oblonga MONTAG, Tab. XX, figg. 850—850 *f.*

Fig. 850: e mari Bahusiæ.

Fig. 850 *c:* apertura coarctata exempli alius.

Figg. 850 *d—850 f:* e sinu Bergensi Norvegico.

Vermiculum oblongum MONT., 1803, Testac. Brit., p. 522, t. 14, f. 9.
Triloc. oblonga D'ORB., 1826, Tab. méth., An. Sc. nat. 7, p. 300, Mod. 95.
Vide præterea GOËS, Ret. Rhizop. Carib. Sea, Sv. Vet. Akad. Handl. 19, 4, p. 124 (ubi Quinquelocul. tenuis CZJZ. est secernenda).
Miliolina oblonga BRADY, 1884, Chall. Rep. Zool. 9, p. 160, t. 5, f. 4.
? Triloc. lævigata (D'ORB.) SCHLUMB., 1893, Mil. golfe Marseille, Mém. Soc. Zool. France 6, p. 205, t. 1, ff. 45—47.

Hab. ad oras Sveciæ occidentales et Norvegiæ profund. metr. 20—50 passim. Long. mm. 1.50.

7. segmentis margine truncatis, aut obtuse carinatis aut rotundate truncatis; segmento antepenultimo projecto angulato aut carinato; testa plerumque opaca, grisea sabulo tenui intermixto constructa; interdum tamen albida subnitida, lævis. Formæ magis inflatæ a M. seminulo difficile distingvendæ:

M. contorta D'ORB., Tab. XX, figg. 851—852.

Fig. 851 *a, b:* e mari Groenlandico profund. metr. 100.

Fig. 851 *c:* magis tumida (interdum tamen submarginata) e mari Atlantico extra Lusitaniam profundo ignoto.

Fig. 852: intermediæ aut angulatæ KARRER proxima e mari extra Väderöarne insulas Bahusiæ profund. metr. 50.

Quinqueloc. contorta D'ORB., 1846, Bass. tert. Vienne, p. 298, t. 20, ff. 4—6.
    »   Juleana, Rodolphina, Badonensis D'ORB., 1846, ibid., p. 298, 299, 300, t. 20, ff.
               1—3, 7—12.
    »   latidorsata RSS, 1849, Neue For. Österr., Wien. Ak. DkSchr. 1, p. 386, t. 50, f. 12.
    »   kostejana, sclerotica KARR., 1868, Mioc. Kostej., Wien. Ak. Sitz. Ber. 58, p. 152,
               t. 3, ff. 4—5.
    »   ? bidentata D'ORB., 1839, For. Cuba, p. 197, t. 12, ff. 18—20.
    »   ? opaca RSS, Sept. Thon. Offenbach, Wien. Ak. S. Ber. 48, p. 42, t. 2, f. 9.
    »   ? Berthelotiana D'ORB., 1839, For. Canaries, p. 142, t. 3, ff. 25—27.
    »   ? ovata (ROEMER), REUSS, 1855, Tert. Sc. nördl. u. mittl. Deutschl., Wien. Ak. S. Ber. 18,
               p. 252, t. 9, f. 88.
Mil. sclerotica BALKW. & MILLETT, 1884, For. Galway, Journ. micr. & nat. Sc. 3, t. 1, f. 2.
? Massilina annectens SCHLUMB., 1893, Mil. golfe Marseille, Mém. Soc. Zool. France 6, p. 220, t. 3,
    ff. 77—79.
? Quinqueloc. rugosa SCHLUMB., ibid., p. 210, t. 4, ff. 91—93.

* gibba:

Triloc. angulata KARR., 1867, Foraminfauna Österreichs Wien. Ak. Sitz. Ber. 55, p. 359, t. 2, f. 6.
    »   intermedia KARR., 1868, Mioc. Kostej., Wien. Ak. Sitz. Ber. 58, p. 138, t. 1, f. 11.

Hab. ad oras occidentales Sveciæ rara metr. 50—70; Groenlandiæ metr. 100; Spetsbergiæ metr. 25.

8. magis applanata, margine plerumque excavato-truncato; a præcedente vix distingvenda:

M. concava REUSS, Tab. XX, fig. 853.

Fig. 853: e Kilsund Norvegiæ profund. metr. 60.

Quinqueloc. concava REUSS, 1849, Neue For. Österr., Wien. Ak. DkSchr. 1, p. 386, t. 51, f. 2.
    »   excavata KARR., 1868, Mioc. For. Fauna Kostej, Wien. Ak. S. Ber. 58, p. 148, t. 2, f. 9.
?   »   bicarinella REUSS, 1869, Oligocän Gass, Wien. Ak. S. Ber. 59, p. 451, t. 1, f. 6.

Hab. in Kilsund Norvegico sinu profund. metr. 60—70 rara.

APPENDIX.

9. segmento antepenultimo etiam truncato valde prominente; interdum obsolete striata, a præcedente vix discernda; opaca:

M. polygona D'ORB., Tab. XX, figg. 854—854 *g.*

Fig. 854 *a, b:* e mari Caraibico profund. metr. 580.
Fig. 854 *c, d:* sectiones transversæ.
Fig. 854 *e, f, g:* obsolete striata, ex eodem loco.

Quinqueloc. polygona D'ORB., 1839, For. Cuba, p. 198, t. 12, ff. 21—23.
? Triloc. quadrilatera D'ORB., ibid., p. 173, t. 9, ff. 14—16.
Mil. seminulum var. Goës, 1882, Ret. Rhizop. Carib. Sea, Sv. Vet. Akad. Handl. 19, 4, t. 9, ff. 353—354.

Hab. mare Caraibicum profund. metr. 500 rara (GOËS). Long. mm. 1.40.

10. segmentis ultimis ambobus truncatis aut emarginatis, singulis costis (aut alis) binis-trinis præditis; segmentis ceteris etiam sæpe carinatis aut alatis:
M. bicostata D'ORB., Tab. XX, fig. 855 *a, b.*
Fig. 855 *a:* e mari Caraibico profund. metr. 500.
Fig. 855 *b:* facies oralis alius, alis paucioribus.

Quinqueloc. bicostata D'ORB., 1839, For. Cuba, p. 195, t. 12, ff. 8—10 (costata, nec ut nostra alata).
Mil. seminulum. var Goës, 1882, Ret. Rhizop. Carib. Sea, Sv. Vet. Akad. Handl. 19, 4, t. 9, ff. 351—352.

Hab. mare Caraibicum profund. metr. 500 rara (GOËS). Long. mm. 1—1.10.

M. secans D'ORB.

Tab. XX, figg. 856—856 *g.*

Plerumque late aut interdum elongate ovalis, complanata, margine attenuato interdum subcarinato, segmentis latis, sæpe transversim undulato-subplicatis.
Fig. 856—856 *d:* e mari Bahusiæ profund. metr. 55.
Fig. 856 *a, e, f:* aperturæ formis variis.
Fig. 856 *g:* sectio transversa.

? Quinqueloc. secans D'ORB., 1826, Tabl. méth., An. Sc. nat. 7, p. 303 (figura Soldanii a D'ORB. citata Mil. bicornem var. angulatam WILLIAMS potius exhibens); Mod. 96.
?   vulgaris D'ORB., ibid., p. 302.
·   Haidingeri D'ORB., 1846, Bass. tert. Vienne, p. 289, t. 18, ff. 13—15.
»   secans ROEM., 1838, Leonh. u. Bronns Jhb. 1838, p. 393, t. 3, f. 77.
Miliolina seminulum v. disciformis WILL., 1858, Rec. For. Gr. Brit., p. 86, t. 7, ff. 188—189.
? Quinqueloc scidula KARR., 1867, For. Österreichs, Wien. Ak. Sitz. Ber. 55, p. 361, t. 3, f. 1.
Miliolina secans BRADY, 1884, Chall. Rep. 9, p. 167, t. 6, ff. 1—2.
Sigmoïlina secans SCHLUMBERG, 1887, Note sur le genre Planispirina, Bull. Soc. Zool. France 12, p. 118.
Massilina secans SCHLUMB., 1893, Mil. golfe Marseille, Mém. Soc. Zool. France 6, p. 218, t. 4, ff. 82—83.

formæ affines:

1. Oblonge ovalis, applanata, obtuse marginata, striolata; striis sæpe valde obsoletis.
M. elegans WILLIAMS, Tab. XX, fig. 857.
Fig. 857: e mari extra Hållö insulam Bahusiæ profund. metr. 50.

Mil. bicornis v. elegans WILL., 1858, Rec. For. Gr. Brit., p. 88, t. 7, f. 195.
? Quinqueloc. Soldanii D'ORB., 1826, Tabl. méth., An. Sc. nat. 7, p. 303.

Hab. ad oras Bahusiæ minus frequens, inter secantem vivens.

2. Inevis aut obsolete ex parte striolata, nunc compressa, nunc subtrigona, margine segmentorum subtruncato, interdum angulis subplicatis; collo interdum paullum producto. Inter Mil. secantem et bicornem medium tenet, cum quibus convivens, a quibus difficillime limitanda; ad M. Ferusacii D'ORB., cui valde propinqua, ab auctoribus forsan sæpe relata.

M. angulata WILL., Tab. XX, fig. 858, Tab. XXI, figg. 859—859 e.

Fig. 858: in secantem vergens, extra Hällö insulam Bahusiæ profund. metr. 50. Tab. XXI, figg. 859—859 e: exempla duo e sinu Bergensium Norvegiæ profund. metr. 18.

Mil. bicornis v. angulata WILL., 1858, Rec. For. Gr. Brit., p. 88, t. 7, f. 196.
Quinqueloc. undosa KARR., 1867, For. Österreichs, Wien. Ak. Sitz. Ber. 55, p. 361, t. 3, f. 3.
? Mil. (Quinqueloc.) Ferusacii PARK. & JONES, 1865, N. Atl. and Arct. Oc., Phil. Trans. 155, tab. 15, f. 36.
? »   undosa BRADY, 1884, Chall. Rep. 9, p. 176, t. 6, ff. 6--8.
? »   seminuda REUSS, 1865, For. deutsch. Sept. Thon., Wien. Ak. DkSchr. 25, p. 125, t. 1, f. 11.

Hab. ad Hällö Bahusiæ, ad Bergen Norvegiæ profund. metr. 20—50 rara. Long. mm. 1.30—1.70.

M. bicornis WALK. & BOYS.

Tab. XXI, figg. 860—861 e.

Ovalis, sectio transversa elliptico-carinata, aut compressa margine truncato, aut trigona, collo nunc producto, nunc abbreviato, apertura nunc circulari nunc triangulari, nunc angusta, elongata; testa plus minusve complete striolata aut subcostata, striolis sæpe undulatis.

Mil. secanti interdum valde propinquans.

Fig. 860, a, b: tenuissime-obsolete striata, var. "eleganti", cnfr. fig. 857, valde propinqua, e Kilsund Norvegiæ profund. metr. 60.

Fig. 861, a, b: ex parte striata, e mari extra Hällö Bahusiæ metr. 50.

Fig. 861, c, d, e: e mari Azorico profund. metr. 80.

Serpula bicornis WALK. & BOYS, 1784, Test. min., p. 1, t. 1, f. 2.
Trilocul. tricostata, Brongniartii D'ORB., 1826, Tab. méth., An. Sc. nat. 7, p. 300; ? Quinqueloc. seminulum, Soldanii D'ORB., ibid. p. 303.
? Triloc. Gualteriana D'ORB., 1839, For. Cuba, p. 170, t. 9, ff. 5—7.
      »   flexuosa D'ORB., 1839, Voy. Amér. mér. 5, p. 73, t. 4, ff. 4—6.
Quinqueloc. nussdorfensis, striatella, obsoleta COSTA, Pal. Nap. 2, p. 326, 328, t. 25, f. 11, t. 26, ff. 3--4.
Mil. bicornis WILLIAMS, 1858, Rec. For. Gr. Brit., p. 87, t. 7, ff. 190—194.
? Triloc. dichotoma RSS, 1849, Neue For. Öster., Wien. Ak. DkSchr., p. 383, t. 49, f. 12.
Quinqueloc. incrassata, Schroeckingeri, vermicularis KARR., 1868, Mioc. Kostej, Wien. Ak. Sitz. Ber. 58, pp. 148—150, t. 2, ff. 10, 12, t. 3, f. 1.
Mil. bicornis BRADY, 1884, Chall. Rep. 9, p. 171, t. 6, ff. 9, 11, 12.
. Quinqueloc. undulata (D'ORB.) SCHLUMB., 1893, Miliol. du Golfe de Marseille, Mém. Soc. Zool. France 6, p. 213, t. 2, ff. 60—61.

Hab. ad Hällö et alia loca Bahusiæ minus frequens profund. metr. 50. Long. mm. 1.80.

formæ affines:

1. forma fere præcedentis, plicato-costata, striis aut costulis inter plicas sæpe inter-
    positis, interdum magis compressa: apertura sæpe circularis; a typo sæpe
    difficillime distingvenda; a Triloc. Linnæana D'ORB., For. Cuba, p. 172, t. 9,
    ff. 11—13, non sat distincta.

M. pulchella D'ORB., Tab. XXI, figg. 862—864.

Figg. 862—863: inter typicam formam et pulchellam D'ORB., e mari Bahusiæ
extra Hållö insulam profund. metr. 50.

Fig. 864: magis typica, e mari Azorico.

Quinqueloc. pulchella D'ORB., 1826, Tab. méth., An. Sc. nat. 7, p. 303. No. 42.
Trilooul. pulchella, Qu. Schreibersii, Josephina D'ORB., 1846, Bass. tert. Vienne, p. 279, 296,
          297, t. 17, ff. 19—21, t. 19, ff. 22—27.
Quinqueloc.   »   BRADY, 1864, Rhizop. Shetl., Transact. Lin. Soc. 24, p. 466, t. 48, f. 1.
       »   »   PARK., JONES, BRADY, 1866, For. Crag., Pal. Soc. 19, p. 13, t. 4, f. 3.
       »   tricarinata D'ORB., 1839, For. Cuba, p. 187, t. 11, ff. 7—9, 13.
       »   plicosa, Josephina COSTA, 1854, Pal. Nap. 2, p. 322, t. 25, ff. 2, 4.
? Mil. paucisulcata RSS, 1864, Oberoligocän, Wien. Ak. Sitz. Ber. 50, p. 452, t. 1, f. 7.
       »   pulchella BRADY, 1884, Chall. Rep., p. 174, t. 3, ff. 10—13, t. 6, ff. 13—14; M. linnæana
               BRADY, ibid. p. 174, t. 6, ff. 15—20 magis complanata cum forma D'ORB.,
               non sat congruens).

Hab. ad Hållö profund. metr. 50 Bahusiæ, rara. Long. mm. 2.

2. magis regulariter constructa et striata aut costulata, a typica non sat distincta.

M. Boueana D'ORB., Tab. XXI, fig. 865.

Fig. 865: e mari Bahusiæ extra Våderöarne insulas profund. metr. 50.

Quinqueloc. Boueana D'ORB., 1846, Bass. tert. Vienne, p. 293, t. 19, ff. 7—9.
       »   affinis, COSTA, 1854, Pal. Nap. 2, p. 329, t. 25, f. 15, t. 25, f. 13.
       »   nussdorfensis D'ORB., 1846, Bass. tert. Vienne, p. 295, t. 19, ff. 13—15.
?   »   Guancha D'ORB., 1839, For. Canar. 143, t. 3, ff. 34—36.
Triloc. Brongniartiana D'ORB., 1839, For. Cuba, p. 176, t. 10, ff. 6—8.
Mil. scrobiculata BRADY, 1884, Chall. Rep. 9, p. 173, t. 113, f. 15.
Quinqueloc. striolata RSS, 1849, Neue For. Österr., Wien. Ak. DkSchr. 1, p. 385, t. 50, f. 10.
       »   costata KARR., 1867, For. Österr., Wien. Ak. Sitz. Ber. 55, p. 362, t. 3, f. 4.
       »   Karreri RSS, 1869, Oligoc. Gaas., Wien. Ak. Sitz. Ber. 59, p. 459.
Triloc. striatella KARR., 1868, Mioc. For. Kostej., Wien. Ak. Sitz. Ber. 58, p. 140, t. 2, f. 2.
       »   porvaensis HKN, 1875, For. Clåv. Szab. Sch., p. 21, t. 13, f. 3.
Mil. boueana BRADY, 1884, Chall. Rep. 9, p. 173, t. 7, f. 13.
? Quinqueloc. disparilis (D'ORB.) SCHLUMBERG. 1893, Mil. golfe Marseille, Mém. Soc. Zool. France 6,
          t. 2, ff. 55—57.

Hab. ad oras Bahusiæ rara, profund. metr. 50. Long. mm. 0.70.

M. tricarinata D'ORB.

Tab. XXI, figg. 866—869.

Triloculina, aliquando qvadriloculina, trigona, interdum elongata, sæpe tamen æque
fere longa ac lata, camerae plerumque magnitudine subæqualibus, suturis angulos acutos
aut carinas (interdum alas) exhibentibus; nitida porcellanea.

Fig. 866: magis elongata, e mari Spetsbergensi profund. metr. 4,630.
Fig. 867: e sinu Gullmaren Bahusiæ profund. metr. 130.
Fig. 868: quadriloculina in seminulum vergens, e mari Caraibico profund. metr. 530.
Fig. 869: procera, e mari Spetsbergensi profund. metr. 350—400.

Triloculina tricarinata D'ORB., 1826, Tab. méth., An. Sc. nat. 7, p. 299, Mod. 94.
Cruciloculina triangularis D'ORB., 1846, Bass. tert. Vienne, p. 280, t. 21, f. 57.
Mil. (Triloc.) **tricarinata** PARK. & JONES, 1865. N. Atl. & Arct. Oc., Phil. Trans. 155, p. 409, t. 15, f. 40.
Triloc. tricarinata PARK., JONES, BRADY, Crag. For., Pal. Soc. 19, p. 7, t. 3, ff. 33—34.
    »   BRADY, 1864, Rhizop. Shetl., Trans. Lin. Soc. 24, p. 466, t. 48, f. 3.
    »   RSS, 1867, Steinsalz Ablag. Wieliczka, Wien. Ak. S. Ber. 55, p. 71, t. 2, f. 4.
Mil. tricarinata BRADY, 1884, Chall. Rep. 9, p. 165, t. 3, f. 17.
 »   »   BRADY, PARK. & JONES. 1887, Abrohlos Bank, Trans. Zool. Soc. Lond. 12, 7, p. 215, t. 40, f. 32.

Hab. ad oras Scandinaviæ metr. 50—180 passim; ad Spetsbergiam metr. 350—4,600. Long. mm. 0.80—5.30.

forma affinis:

 cameris paullum inflatis, angulis suturarum minus acutis aut rotundato-obtusis; in
Mil. trigonulam transiens:

M. gibba D'ORB., 1846, Bass. tert. Vienne, p. 274, t. 16, ff. 22—24.
Mil. gibba EGGER, 1857, Mioc. Ortenburg, Leonh. u. Bronns Jhb. 1857, p. 271, t. 6, ff. 1—3.
 »   »   v. IlKN, 1875, Clávul. Szaboi Sch., p. 21, t. 12, f. 10.

M. trigonula LMCK.

Tab. XXII, fig. 870.

 Sphærica aut ovata, trigona, inflata, angulis rotundatis, suturis sæpe callosis; sæpe
subgrisea minus nitens, subarenosa; a Mil. gibba difficillime distincta, a formis Miliol.
ringentis juvenilibus sive triloculinis sæpe vix limitanda.
 Fig. 870: e Gullmaren sinu Bahusiæ profund. metr. 50.

Miliolites trigonula LMCK, 1804, Ann. d. Mus. 5, p. 351.
 » **cor angvinum** BLAINV., 1825, Man. d. Malac. et Conch., p. 369, t. 4, f. 3.
Triloc. trigonula D'ORB., 1826, Tab. meth., Ann. Sc. nat. 7, p. 299, Mod. 93.
 » **austriaca** D'ORB., 1846, Bass. tert. Vienne, p. 275, t. 16, ff. 25—27.
? **Miliol. austriaca** EGGER, 1857, Mioc. Ortenburg; Leonh. u. Bronns Jhb. 1875, p. 271, t. 6, ff. 4—6.
 » **trigonula** WILLIAMS, 1858, Rec. For. Gr. Brit., p. 83, t. 7, ff. 180 182.
Triloc. gibba RSS, 1864, Oberoligocän, Wien. Ak. Sitz. Ber. 50, p. 450, t. 1, f. 4.
Mil. trigonula BRADY, 1884, Chall. Rep. 9, p. 164, t. 3, ff. 15—16.

Hab. ad oras Scandinaviæ profund. metr. 50—180 semper Mil. simplicem et varietates hujus concomitans,
minus frequens. Long. mm. 0.90.

M. valvularis BRADY.

Tab. XXII, fig. 871.

 Plus minusve obsolete tri-quadriloculina, ovalis ex margine plerumque compressa,
nitida, lævis, cameris externis 2—3, sequente totam aut ex parte præcedentem amplectente,
quare testa plus minusve incrassata; apertura rima semicircularis interdum crenata.

Facies externa et modus testæ accrescendi in memoriam revocant Mil. sigmoideam BRADY.

Fig. 871: facies segmenti penultimi; *a:* facies oralis; *b:* facies marginalis; *c:* facies segm. penultimi alius magis compressa; *d:* sectio transversa; *e* mari Atlantico boreali profund. metr. 1,740.

Mil. valvularis BRADY, 1884, Chall. Rep. 9, p. 161, t. 4, ff. 4—5.

Obs. Triloculina valvularis Rss, 1851, Sept. Thon. Berlin, Zeitschr. deut. geol. Gesellsch. 3, p. 85, t. 7. f. 56, a Trilor. cnoplostoma et turgida Rss, ibid., p. 86, t. 7, ff. 57—58, et in SCHLICHTI, Septuricuth. Pietzpuhl, Wien. Ak. S. Ber. 62, t. 36, ff. 4—17, non vere distincta, verosimiliter ut varietas triloculina & quadriloculina Mil. simplicis. D'ORB., habenda, quare nomen "valvularis" Rss, delendum est. Præterea hæc nomen in lævigatam BORNEMANN, 1855, Sept. Thon. Hermsdorf; Zeitschr. deutsch. geol. Gesellsch. 7, p. 350, t. 19, f. 5 transferri potest.

Hab. mare Atlanticum boreale metr. 1,740 rara (LINDAHL). Long. mm. 1.

** Biloculina.

B. bulloides D'ORB.

Nitida, sphæroidea aut ovoidea, plerumque haud marginata aut margine segmenti ultimi obtuso, interdum subacuto; apertura fere circulari, lingua plerumque bifurcata aut semilunari munita.

Bil. bulloides D'ORB., 1826, Tab. méth., An. Sc. nat. 7, p. 297, t. 16, ff. 1—4, Mod. 90.
»   »   SCHLUMBERGER, 1887, Note sur les Biloc. bulloides D'ORB. et Bil. ringens LAM., Bullet. Soc. geol. France (3) 15, p. 119, t. 15, ff. 10—13.

Obs. Forma **ringens** LAM. a Cel. Dom. SCHLUMB. sic est distingvenda: ovoidea, non depressa, adultæ sparsim transverse subplicatæ, margine rotundato aut obsoleto, apertura fere circulari, grandi, testæ partem crassitudinis tertiam æquante, limbata, lingua bifurcata, aut stylo biauriculato, auriculis perforatis, aut lamina obovata, pedunculata foramine centrali perforata, munita.
SCHLUMB., ibid., p. 126, tab. 15, ff. 14—18.
Forma utraque extincta sec. SCHLUMBERGER.

formæ affines:

1. plerumque obtuse marginata, apertura ovata aut rima semicirculari, aut fibulæformi, nunc brevi nunc valde elongata, præterea typicæ similis, a qua non sat distincta. Margo posticus interdum denticulatus, plicatus aut incrassatus. Haud rare triloculina, id est stadium larvale continuatum.
B. simplex D'ORB., Tab. XXII, figg. 872—882, 885, Tab. XXIII, figg. 886—887.

Fig. 872: apertura magna, lingua bicornuta.
Fig. 873: apertura parva, lingua cruciformi; ambæ e fretis Koster insulam profund. metr. 35—132.
Fig. 874 *a, b, c:* B. inornatæ D'ORB. similis; exempla duo e Gullmaren sinu Bahusiæ profund. metr. 60.
Fig. 875: forma triloculina, apertura fibulæformi longa, ex eodem loco.
Fig. 876: triloculina, magis elongata, e sinu Bergensium Norvegiæ profund. metr. 125.
Fig. 877: ex eodem loco.

Fig. 878: triloculina, e Hækkefjord Norvegiæ, forsan ad formam abyssorum referenda.

Figg. 879—880: sectiones transversæ duorum exemplorum, dimorphismum larvale exhibentes; e mari Bahusico.

Fig. 881: "Bil. appendiculata" Reuss, e mari Spetsberg. profund. metr. 80.

Fig. 882: exemplum duplicatum, aperturis formis interse discrepantibus.

Fig. 885: in depressam vergens, e mari Spetsberg. profund. metr. 50; B. brachyodontæ Fornas. propinqua.

Tab. XXIII, fig. 886: facies oralis formæ magis tumidæ, e mari Azorico profund. metr. 530—1,060; a: subtriloculina, ex eodem loco.

Fig. 887: triloculina ex eodem loco (forma larvalis).

Biloc. simplex, clypeata d'ORB., 1846, Bass. tert. Vienne, pp. 263, 264, t. 15, ff. 19—21, 25—27.
> bulloides, simplex, circumclausa COSTA, 1854, Pal. Nap. 2, p. 299, 300, 307, t. 24, ff. 1, 3, 6.
> turgida Rss, 1851, Septar. Thon. Berlin, Ztschr. deut. geol. Gesellsch. 3, p. 85, t. 7, f. 55; v. SCHLICHT, Sept. Thon. Pietzpuhl, t. 35, ff. 27—29, t. 36, ff. 1—3.
> ringens WILLIAMS, 1858, Rec. For. Gr. Brit., p. 79, t. 6, 7, ff. 169—171.
> bulloides v. calostoma, anodonta KARR., 1868, For. Kostej., Wien. Ak. S. Ber. 58, p. 133, t. 1, ff. 4, 6.
> globiformis KARR., 1867, For. Österreichs, Wien. Ak. S. Ber. 55, p. 357, t. 2, f. 1.
> canariensis d'ORB., 1839, For. Canaries, p. 139, t. 3, ff. 10—12 (in elongatam d'ORB. vergens.)
> affinis, inornata d'ORB., 1846, Bass. tert. Vienne, p. 265 266, t. 16, ff. 1—3, 7—9 (in elongatam vergens.)
> caudata BORNEM., 1855, Sept. Thon. Hermsd., Zeitschr. deut. geol. Gesellsch., p. 348, t. 19, f. 2.
> ringens PARK., JONES, BRADY, 1866, For. Crag., Pal. Soc. 19, p. 5, t. 3, ff. 26—28.
> serrata BAILEY, 1862, Notes new spec. micr. obj., Boston Journ. Nat. Hist. 7, p. 350, t. 8.
Triloc. bipartita d'ORB., 1846, Bass. tert. Vienne, p. 275, t. 17, ff. 1—3.
> enoplostoma REUSS, 1851, Sept. Thon., Berl., Zeitsch. deutsch. geol. Gesellsch. 3, p. 86, t. 7, f. 57; v. SCHLICHT, 1870, Sept. Thon. Pietzpuhl, t. 36, ff. 4—17 (forma larvalis).
Biloc. ringens BRADY, 1884, Chall. Rep. 9, p. 142, t. 2, ff. 7—8.
? Biloc. intermedia FORNASINI, 1886, Bollet. Soc. geol. Ital. 5, t. 4, f. 2; 1893, Mem. Sc. Instit. R. Bologna (5) 3, t. 1, f. 1.
ringens BRADY, PARK., JONES, 1887, Abrohlos Bank, Trans. Zool. Soc. Lond. 12, 7, p. 213, t. 40, ff. 19—20.

Hab. ad oras Scandinaviæ occidentalis profund. metr. 30—150 nec non mare Spetsbergicum et Groenlandicum profund. metr. 50—200 haud raru.

2. striolata aut semistriolata, apertura plerumque brevi.

B. comata BRADY, Tab. XXII, figg. 883—884.

Fig. 883: semistriolata, e fretis Koster insularum profund. metr. 133.

Fig. 884: apertura lingua biauriculata valde prominente, e Gullmaren sinu Bahusiæ profund. metr. 130.

Biloc. comata BRADY, 1884, Chall. Rep. 9, p. 144, t. 3, f. 9.
> > SCHLUMBERG., 1891, Biloc. grands fonds, Soc. Zool. France 4, p. 178, t. 10, ff. 72—73.

Hab. in mari Bahusico rara.

3. inflata, lævis, e margine plerumque paullum compressa, i. e. altior paullo quam latior, non marginata, apertura nunc rima semilunari nunc fissura longa angulosa, irregulari. Bil. globulo Rss forsan propinqua.

B. abyssorum n., Tab. XXIII, figg. 888—889.

Fig. 888: exemplum ex Atlantico boreali profund. metr. 1,750.

Fig. 889: e mari Spetsbergensi profund. metr. 500.

Hab. mare Arcticum et boreale profund. metr. 500—2,000; mare Azoricum profund. metr. 700.

4. nitida, plus minusve irregulariter tri-quinqueloculina, globosa aut ovalis ex margine saepe paullum compressa; apertura rima semicirculari aut angulari aut irregulariter undata, interdum longissima, supramarginali; segmento antepenultimo nunc laterali nunc frontali, Forma multiloculina praecedentis.

M. bucculenta BRADY, Tab. XXIII, figg. 890—903; Tab. XXIV, figg. 904—905.

Fig. 890: juvenis, e mari Atlant. boreali profund. metr. 1,740.

Fig. 891: Triloc. gibbae D'ORB. similis, ex eodem loco.

Fig. 892: Triloc. enoplostomae RSS similis; ex eodem loco.

Fig. 893: ? Juvenis, e mari Spetsbergensi, inter M. bucculentam, profund. metr. 1,388.

Fig. 894: ex eodem loco.

Fig. 895: Triloc. trigonulae LAMCK. similis, e mari Azorico profund. metr. 530.

Figg. 896—901: formae variae, tri-quadriloculinae, e mari Azorico profund. metr. 530—1,060.

Fig. 902: quadriloculina, e mari Spetsbergensi, profund. metr. 350.

Fig. 903: e mari Atlantico boreali profund. metr. 1,740.

T. XXIV, fig. 904: quadriloculina, facies lateralis; a: facies ventralis sive frontalis, e mari Spetsbergico profund. metr. 450.

Fig. 905: triloculina, facies lateralis, ex eodem loco.

BRADY, Chall. Rep., 1884, p. 170, t. 114, f. 3.
M. bucculenta v. placentiformis BRADY, ibid., p. 171, t. 4, ff. 1—2.
M. cryptella PARK. & JONES, 1865; N. Atl. & Arct. Oc., Phil. Trans. 155, p. 410, t. 15, f. 39.
M. ringens var. GOËS, 1882, Rct. Rhizop. Carib. Sea, Sv. Vet. Akad. Handl. 19, 4, t. 10, ff. 374—375.
Planispirina bucculenta SCHLMB., 1892, Mém. Soc. Zool. France 5, p. 194.

Hab. mare Arct. glaciale profund. metr. 350—1,400 passim (v. OTTER). Long. mm. 2.50; mare Azoricum (SMITT & LJUNGMAN).

5. elliptica, sublagenaeformis, impolita saepe in collum breve producta, apertura circularis; testa saepe albidogrisea, e pellicula arenosa tenui subscabra; saepe tri-sexloculina.

B. tubulosa COSTA.
1854, Pal. Nap. 2, p. 309, t. 24, f. 7.
Biloc. lucernula SCHW., 1866, For. Kar. Nikob., Novara Reise geol. Th. 2, p. 202, t. 4, ff. 14, 17.
Mil. ringens var. GOËS, 1882, Rct. Rhizop. Carib. Sea, Sv. Vet. Akad. Handl. 19, 4, t. 10, ff. 363—365, 376—383.
Biloc. bulloides BRADY, 1884, Chall. Rep. 9, p. 142, t. 2, ff. 5—6.
»   **tubulosa** BRADY, ibid., p. 147, t. 3, f. 6 (triloculina).
»   **lucernula** SCHLUMBERG. 1891, Bilocul. des grands fonds, Mém. Soc. Zool. France 4, p. 185, t. 12, ff. 90—96.

Hab. mare Atlanticum metr. 530—2,000 passim. Diam. mm. 2.40.

6. magis angusta, elongate ovalis aut elliptica, segmento penultimo sæpe postice attenuato, interdum fimbriato aut caudato ultimum excedente; margine plerumque obtuso aut rotundato; a typica non limitanda; nunc teres tumida, nunc depressa; formæ tri-quadriloculinæ etiam occurrunt.

B. elongata d'Orb., Tab. XXIV, figg. 906—913.

Fig. 906: in typicam vergens, e Hækkefjord, Norvegiæ, metr. 180.

Figg. 907—913: formæ bi-, tri-, quadriloculinæ, aliquot in Biloc. simplicem vergentes; omnes a mari Bahusiæ Sveciæ, profund. metr. 30—65.

1826. Tab. méth., An. Sc. nat. 7, p. 298, No. 4.
Biloc. oblonga d'Orb., 1839, For. Cuba, p. 163, t. 8, ff. 21—23.
» **Patagonica, Bougainvillei** d'Orb., 1839, Voy. Amér. mér. 5, p.65,67,t.3,ff.15—17,t.8,ff.22—24.
» **constricta** Costa, 1854, Pal. Nap. 2, p. 301, t. 24, f. 2.
» **ringens** var. **patagonica** Williams, 1858, Rec. For. Gr. Brit., p. 80, t. 7, ff. 175—176.
Mil. (Biloc.) elongata Park. & Jones, 1865, N. Atl. & Arct. Oc., Phil. Trans. 155, p. 409, t. 17, ff. 88, 90—91.
Biloc. bulloides v. **truncata, gracilis** Rss, 1867, Steinsalzablag. Wieliczka, Wien. Ak. Sitz. Ber. 55, p. 68, t. 2, ff. 1—2.
» **tenuis** Karr., 1868, Mioc. Kostej, Wien. Ak. Sitz. Ber. 58, p. 133, t. 1, f. 5.
» **elongata** Brady, Chall. Rep. 9, p. 144, t. 2, f. 9.
» » Brady, Park. & Jones, 1887, Abrohlos Bank, Trans.Lin.Soc.12, 7, p.214, t.40, ff.21—22.
» » Schlumb., 1891, Biloc. grands fonds, Mém. Soc. Zool. France 4, p.184, t.11, 12, ff.87—89.

Hab. ad oras Scandinaviæ occidentales cum typica passim.

7. ovalis, subdepressa, segmentis ultimis margine subcarinatis. Segmentum ultimum postice sæpe incrassatum et emarginatum. Inter typicam et var. depressam medium tenens, a neutra limitanda.

B. lævis Defr., Tab. XXIV, figg. 914—918.

Figg. 914—917: formæ variæ e mari Spetsbergico profund. metr. 1,450—4,600.

Pyrgo lævis Defr., 1824, Dict. Sc. nat. 32, p. 273, t. 88, f. 2.
Biloc. lævis d'Orb., 1826, Tab. méth., An. Sc. nat. 7, p. 298, No. 8.
» » Brady, 1884, Chall. Rep. 9, p. 146, t. 2, ff. 13—14.
» **appendiculata** Rss, 1863, Crag d'Anvers, Bull. Acad. Belg. (2) 15, p. 139, t. 1, f. 1.
» **amphiconica** var. **platystoma** Rss, 1867, Steinsalzablag. Wieliczka, Wien. Ak. Sitz. Ber. 55, p. 67, t. 1, f. 8.
Mil. ringens var. Goës, 1882, Ret. Rhizop. Carib. Sea, Sv. Vet. Akad. Handl. 19, 4, t. 10, ff. 361—362.

Hab. mare Arcticum metr. 1,450—4,600 vulgaris (v. Otter).

8. nitida globosa aut ovalis aut subdepressa, segmento ultimo penultimum valde excedente sæpe partim amplectente; apertura extensa, sæpe rima protracta, angulata aut undulata, suprasuturalis; in var. depressam d'Orb. transiens.

B. arctica Goës, Tab. XXV, figg. 919—920.

Fig. 919: procera, e mari Spetsbergensi profund. 4,600.

Fig. 920: in var. abyssorum vergens; ex eodem mari, profund. metr. 350.

Mil. (Biloc.) ringens Park. & Jones, 1865, N. Atl. & Arct. Oc., Phil. Trans. 155, p. 409, t. 15, ff. 42—43.
? **Biloc. larvata** Rss. 1867, Steinsalzablag. Wieliczka, Wien. Ak. S. Ber. 55, p. 70, t. 2, f. 3.

Hab. mare Arcticum profund. metr. 350—1,400 passim Biloc. depressam concomitans (v. Otter). Long. mm. 2—3.

B. depressa d'Orb.

Tab. XXV, figg. 921—925.

Nitida, orbicularis aut ovalis, depressa, segmento ultimo (et interdum penultimo) carinato-marginato; apertura rima protracta transversa ant fibuleeformnis ant subcircularis, lingula bifurcata sæpe provisa; margo posticus interdum emarginatus. A B. simplici non sat limitata.

Fig. 921: ad "serratam" tendens, e mari Azorico, profund. metr. 980.

Fig. 922: e Skagerack, profund. metr. 270.

Fig. 923: sectio transversa, stadium larvale Vloculinum exhibens; e mari Spetsbergensi, profund. metr. 4,600; in B. simplicem vergens.

Fig. 924—925: e mari Bahusiæ profund. metr. 35—40.

1826, Tab. méth., Ann. Sc. nat. 7, p. 298, No. 7, Mod. 91.
Biloc. carinata D'ORB., 1839, For. Cuba, p. 164, t. 8, f. 24; t. 9, ff. 1—2.
»   **lunula** D'ORB., 1846, Bass. tert. Vienne, p. 264, t. 15, ff. 22—24.
»   **amphiconica** Rss, 1849, Neue For. Österr., Wien. Ak. DkSchr., p. 382, t. 49, f. 5.
»   **ringens** var. **carinata** WILL., 1858, Rec. For. Gr. Brit., p. 79, t. 7, ff. 172—174.
Mil. (Biloc.) depressa PARK. & JONES, 1865, N. Atl. & Arct. Oe., Phil. Trans. 155, p. 409, t. 17, f. 89.
Biloc. depressa JONES, PARK., BRADY, 1866, For. Crag., Pal. Soc. 19, p. 6, t. 3, ff. 29—30.
»   **scutella** KARR., 1868, Mioc. Kostej., Wien. Ak. Sitz. Ber. 58, p. 134, t. 1, f. 7.
»   **plana** KARR., 1877, Hochqu. Wasserleit., Abh. geol. Reichsanst. 9, p. 375, t. 16, f. 9.
»   **ringens** var. GoËs, 1882, Ret. Rhizop. Carib. Sea, Sv. Vet. Akad. Handl. 19, 4, t. 10, ff. 366—367, (in serratam vergens).
»   **depressa** BRADY, 1884, Chall. Rep. 9, p. 145, t. 2, ff. 12, 15—17; t. 3, ff. 1—2.
»   »   BRADY, PARK., JONES, 1887, Abrolhos Bank, Trans. Zool. Soc. Lond. 12, 7, p. 213, t. 40, ff. 17—18.
»   »   SCHLUMBERG. 1891, Biloc. grands fonds, Mém. Soc. Zool. France 4, p. 160, t. 9, ff. 48—49.

Hab. ad oras occidentales Scandinaviæ metr. 40—500 sat frequens, mare Spetsbergicum metr. 900—1,800, haud rara. Diam. mm. usque ad 2.10.

forma affinis:

margine segmentorum serrato-crenato, apertura plerumque ovali aut subcirculari; a typica non sat limitata.

B. serrata BRADY, Tab. XXV, fig. 926.

Fig. 926: e mari Atlantico boreali profund. metr. 1,740.

Biloc. depressa var. **serrata** BRADY, 1884, Chall. Rep. 9, p. 146, t. 3, f. 1.
»   **serrata** SCHLUMBERG. 1883, Feuille jeunes natural. (separ.) 13, p. 23, t. 3, f. 3.
»   »   SCHLUMBERG. 1891, Bil. grands fonds, Mém. Soc. Zool. France 4, p. 163, t. 9, ff. 50—51.

Hab. mare Atlanticum boreale profund. metr. 1,740.

B. sphæra d'Orb.

Tab. XXV, fig. 927.

Globosa, nitidissima, segmento ultimo valido, magnam penultimi vix prominentis partem amplectente; apertura nunc regularis rima semicircularis aut angulata aut rimæ undatæ sæpe ramosæ.

Fig. 927: e fretis Koster insulam profund. metr. 70—80.

Biloc. sphæra n'Orb., 1839, Voy. Amér. mérid. 5, p. 66, t. 8, ff. 13—16.
? » globulus (v. Schlicht) Reuss, 1870, Sitz.Ber. Wien. Ak. 62, t. 464; v. Schlicht, tab. 35, ff. 30—32.
? » » Bornem., 1855, Sept. Thon. Hermsdorf, Zeitschr. deut. geol. Ges. 7, p. 349, t. 19, f. 3.
ringens var. Goës, 1882, Ret. Rhizop. Carib. Sea, Sv. Vet. Akad. Handl. 19, 4, t. 10, ff. 368—369.
sphæra Brady, 1884, Chall. Rep. 9, p. 141, t. 2, f. 4.
» sphæroidea Schlumb., 1880, Fenille Jennes Natur. 13, p. 22, t. 2, f. 3, a nostra non sat distincta.
Planispirina sphæra Schlumberg, 1891, Biloc. grands fonds, Mém. Soc. Zool. France 4, p. 190, ff. 45—46.

Hab. ad oras Scandinaviæ occidentales et Groenlandiæ profund. metr. 70—1,250 minus frequens.

APPENDIX.

B. saccata n.

Tab. XXV, fig. 928.

Biloculina a margine compressa, haud polita, segmento ultimo lagenæformi, sæpe collo tubulari instructo; segm. penultimo magis compresso fere toto nudo, aut distante segmenti partem antepenultimi non obtegente: apertura circularis, lingula plerumque bifurcata.

Ut B. tubulosa Costa constructa. tamen compressa et forma alia.

Fig. 928: exemplum ordinarium cum facie orali a.

b: exemplum hians, segmentum antepenultimum præbens.

c: sectio transversa.

d: exempl. aliud valde compressum, collo abreviato; omnia e mari Caraibico.

Hab. mare Caraibicum profund. metr. 500 rara (Goës). Long. mm. 1.10.

B. quadrangularis n.

Tab. XXV, fig. 929.

Biloculina, ovalis, tetragona, subæquilateralis, segmentis subtriangularibus, ultimo (interdum et penultimo) marginato-carinato, angulo dorsali amborum uni-bicarinato, carinis sæpe undulato-crenatis, imcompletis; apertura rima suturalis circumflexa.

Fig. 929: facies segmenti penultimi.

a: facies oralis.

b: facies marginalis.

c: sectio transversa; e mari Caraibico.

Inter Biloculinas nulla est relata forma angulosa nisi Biloc. aculeata d'Orb., An. Sc. nat. 7, p. 298, No. 3, Mod. 31, ex Aquitania Galliæ australis, fossilis.

Goës, 1882, Ret. Rhizop. Carib. Sea, Sv. Vet. Akad. Handl., t. 9, ff. 348—349.

Hab. mare Caraibicum profund. metr. 500 rara (Goës). Long. mm. 1.

CERATINA n. g.
C. trochamminoides n.
Tab. XXV, fig. 930.

Plus minusve regulariter planospiralis, haud polita, margine rotundato. Anfractus subteretes, apparentes 3—4—5, interni sæpe irregulariter convoluti, segmentis ultimi anfr. 10—12, suturis incisis, septis plus minusve completis.

Exemplorum paucitatis causa hoc genus sat investigare non potui, Trochamminæ proteo KARR. similis, structura certe tamen porcellanea.

Fig. 930: facies lateralis.

a: facies marginalis.

b: septum fere completum.

c: sectio transversa, stadium larvale miliolinæforme præbens.

Hab. Azores insulas mare alluens profund. metr. 540 (SMITT, LJUNGMAN).   Diam. mm. 1.15.

INDEX.

EXPOSITIO TABULARUM.

(Notæ numerorum tenues mensuram indicant millimetricam.)

TAB. I.

Tab. I.

Figg. 1—3: Astrorhiza limicola SANDAHL...

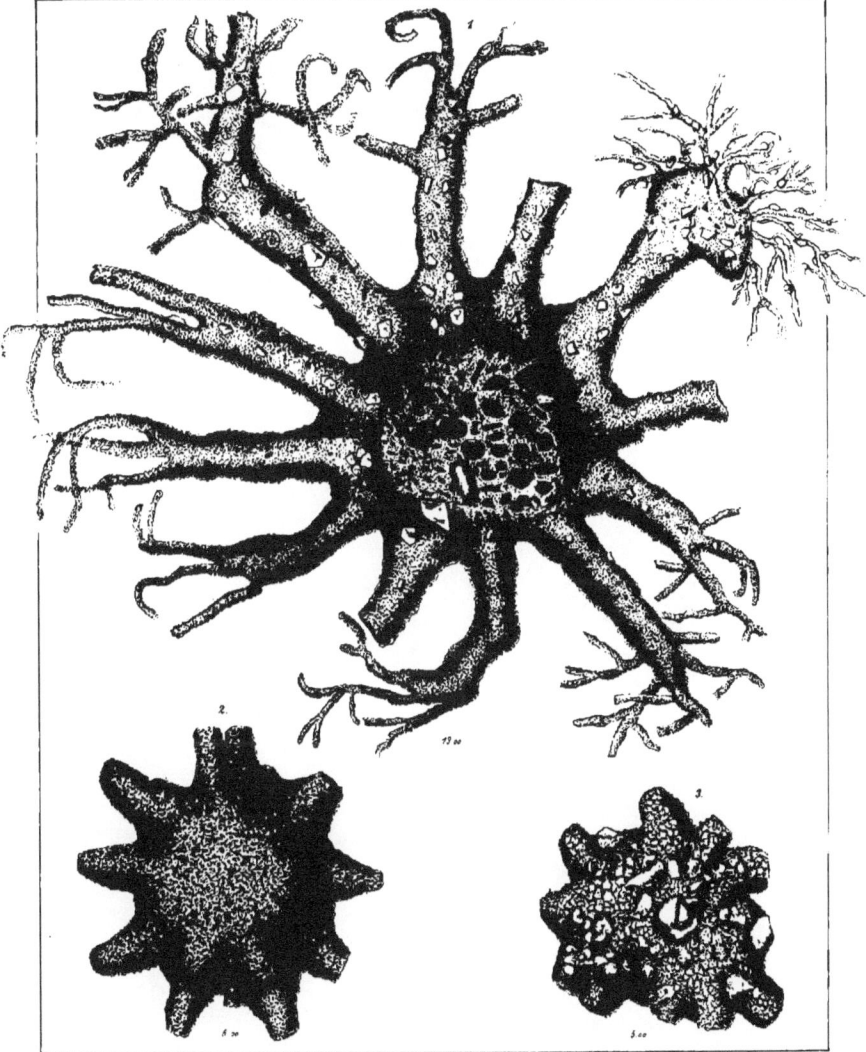

TAB. II.

Tab. II

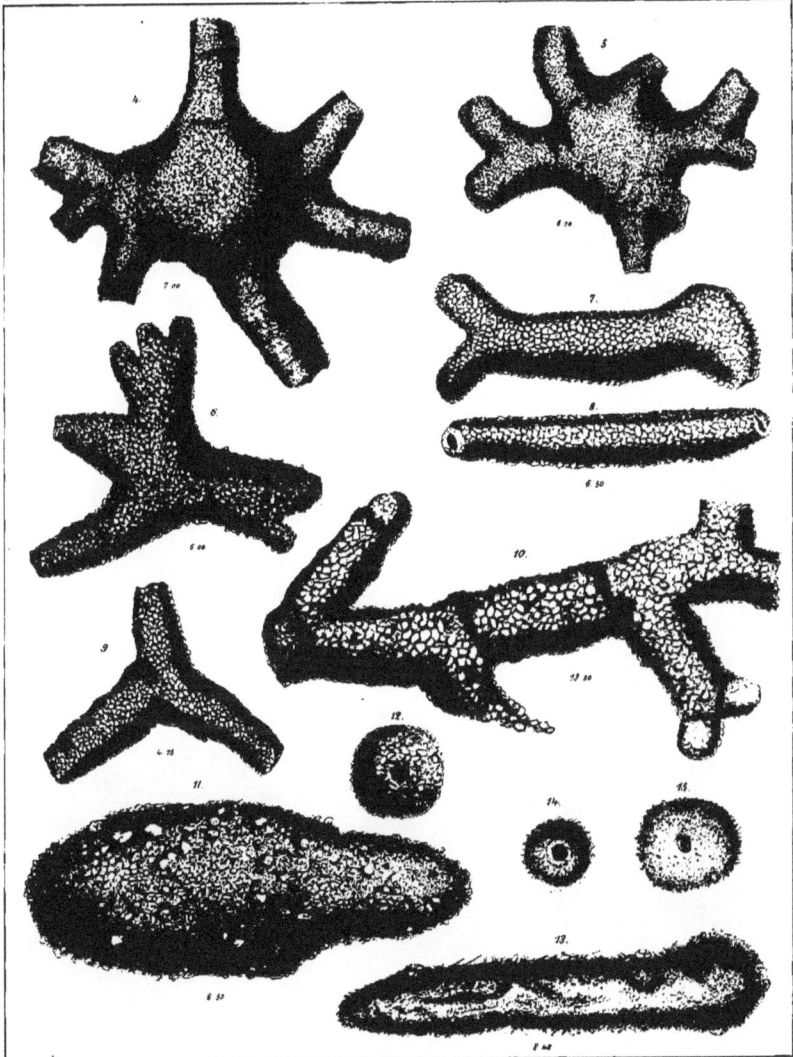

TAB. III.

Tab. III.

Kgl. Vet. Akad! Handl. Bd 25. N? 9.

Tab. III

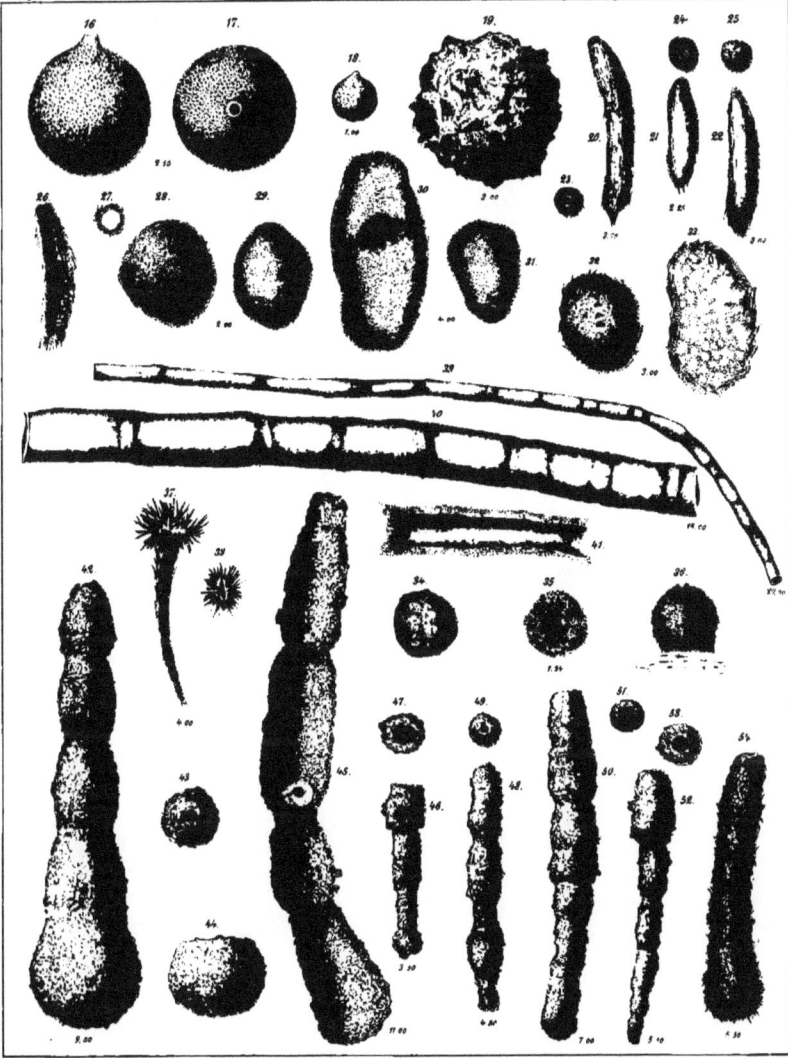

A. Goës & Hedelin del

G. Tholander. lith.

W. Schlachter Stockholm.

TAB. IV.

Tab. IV.

Tab. IV

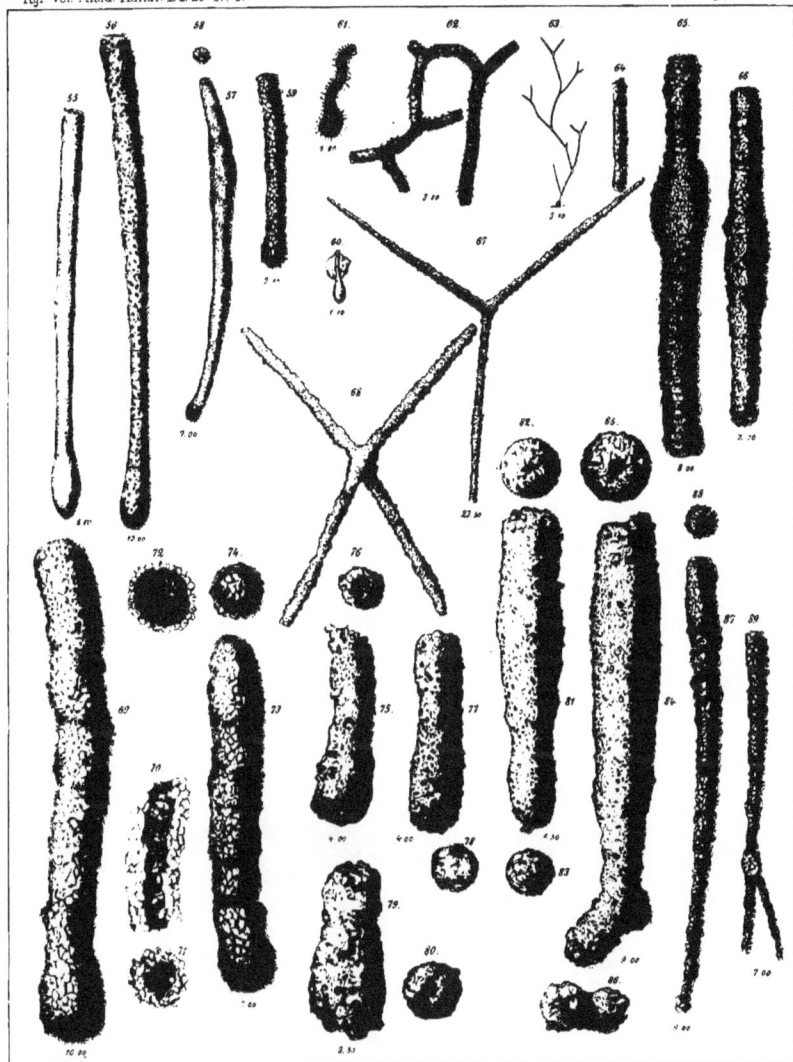

Tab. V.

Tab V

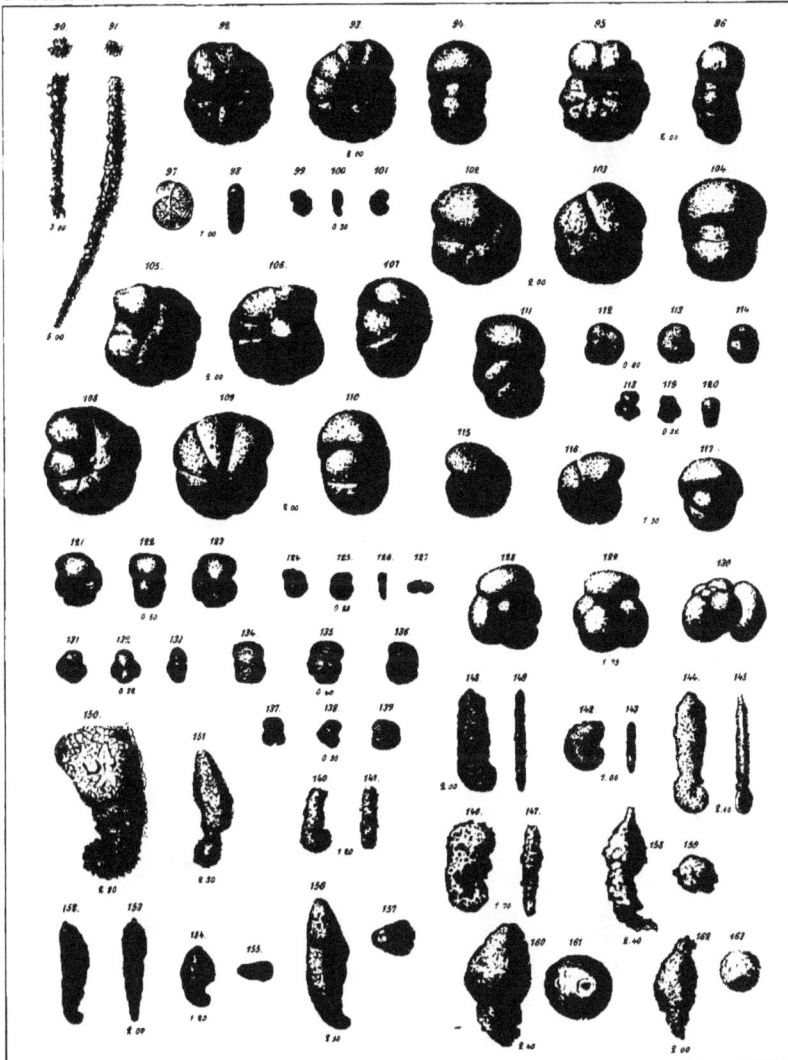

TAB. VI.

Tab. VI.

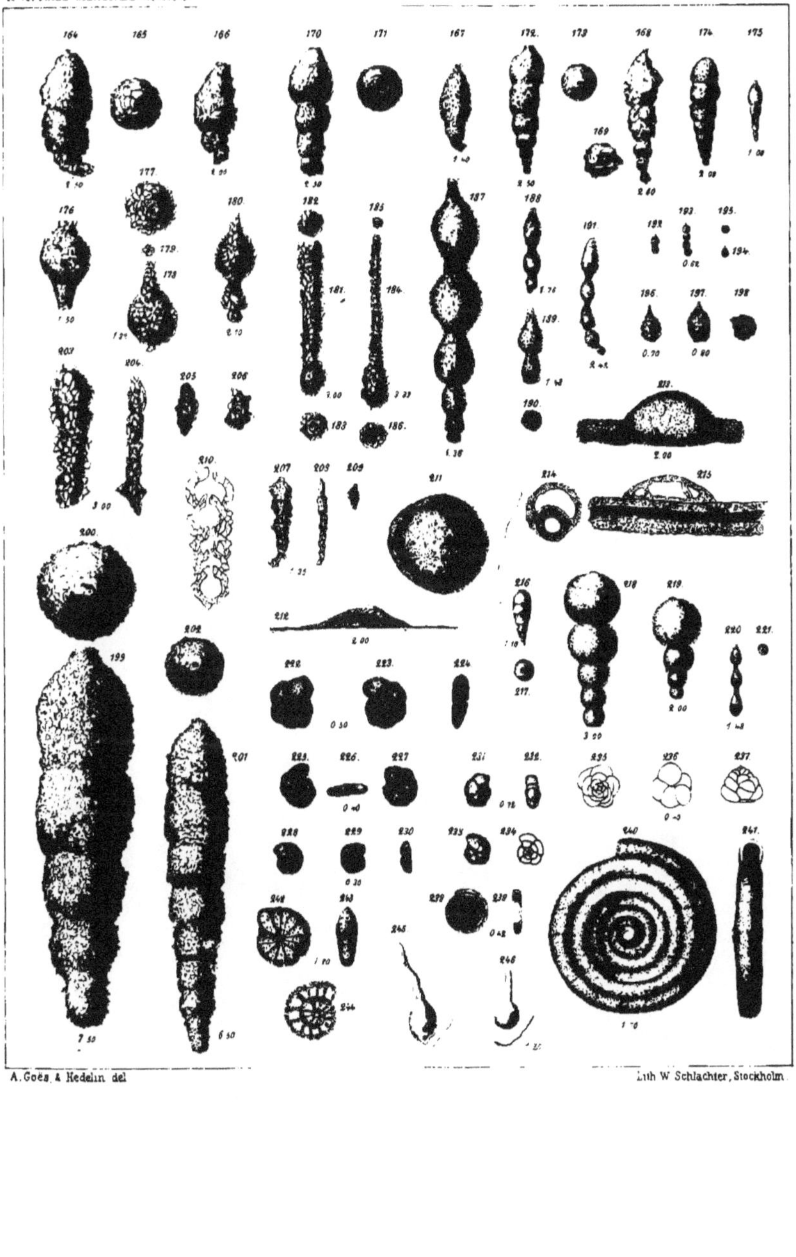

A. Goës & Hedelin del Lith W Schlachter, Stockholm.

TAB. VII.

Tab. VII.

Tab. VI.

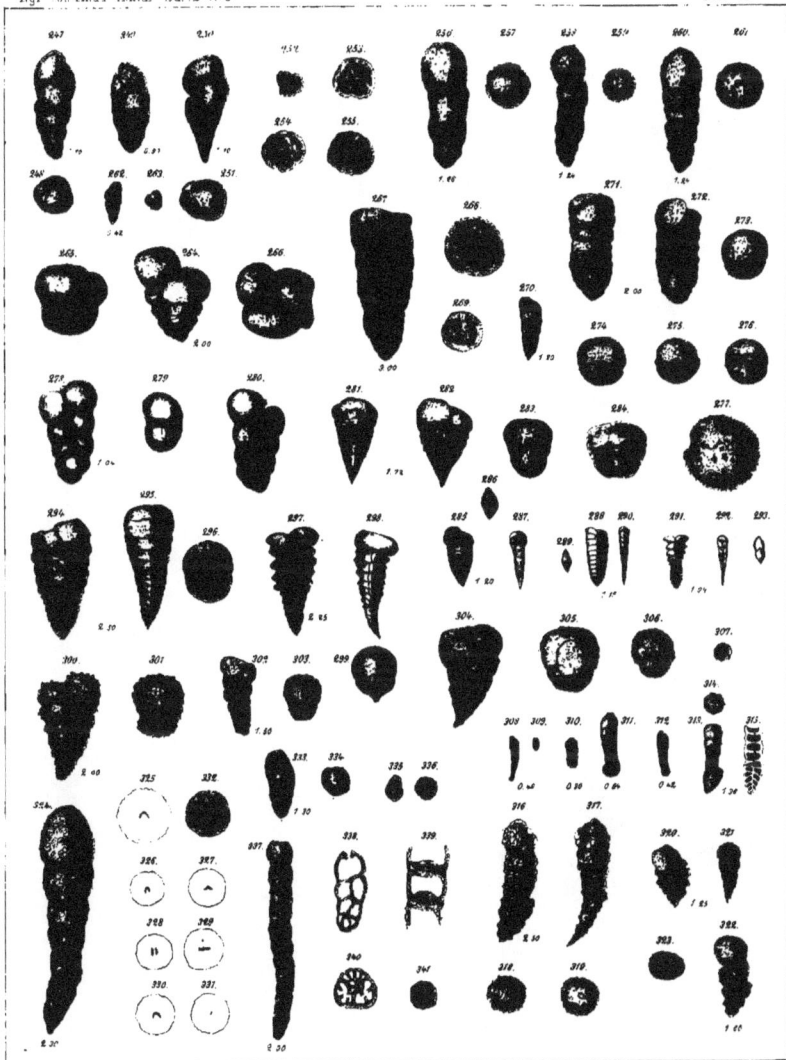

TAB. VIII.

———

Tab. VIII.

Tab. VIII.

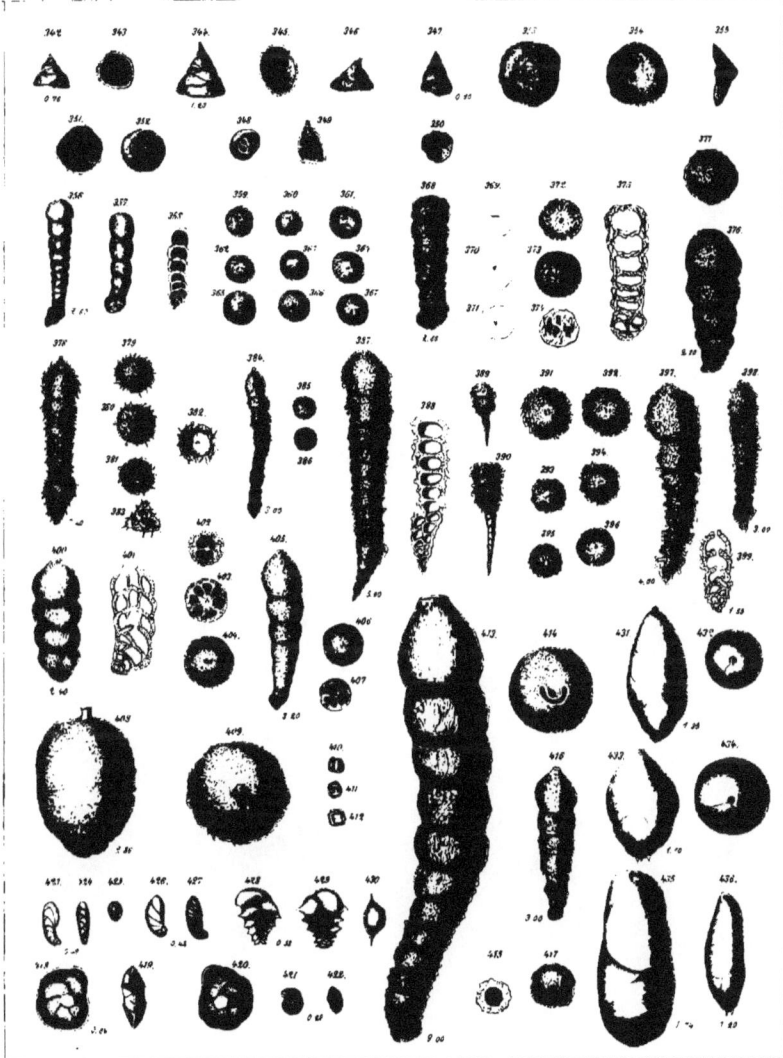

A. Goës del.
O. Tholander. lith
W. Schlachter Stockholm.

TAB. IX.

Tab. IX.

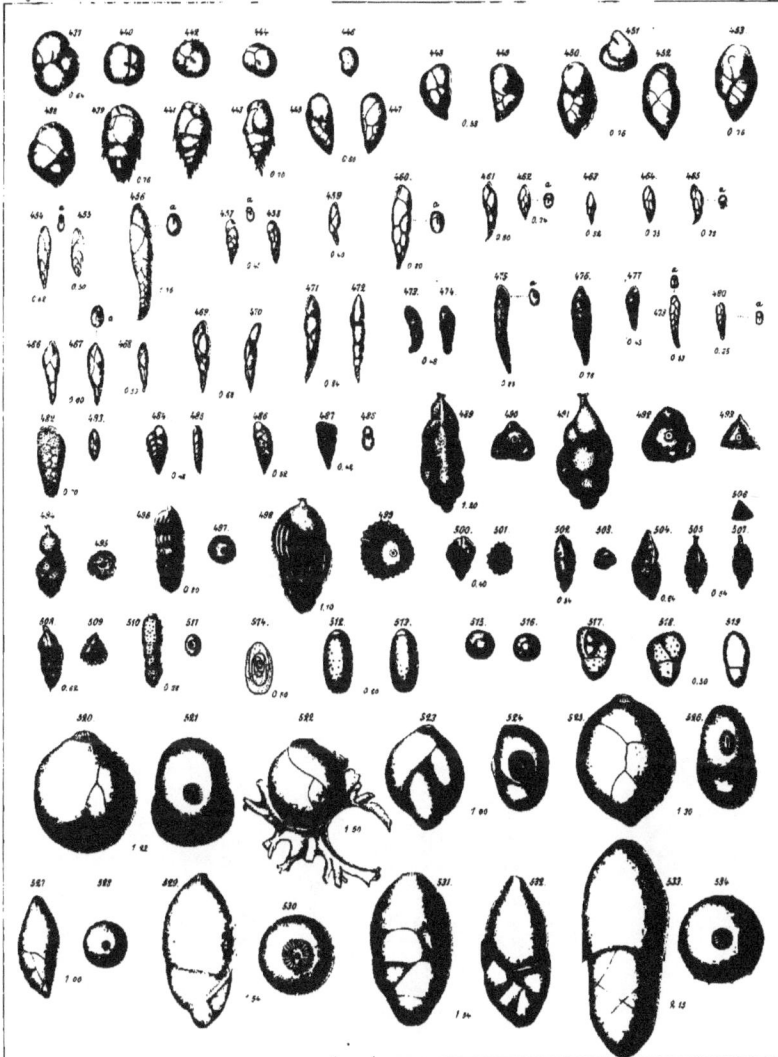

 Tab. IX

A. Goës del. G. Thoïander lith. W. Schlachter, Stockholm.

TAB. X.

Tab. X.

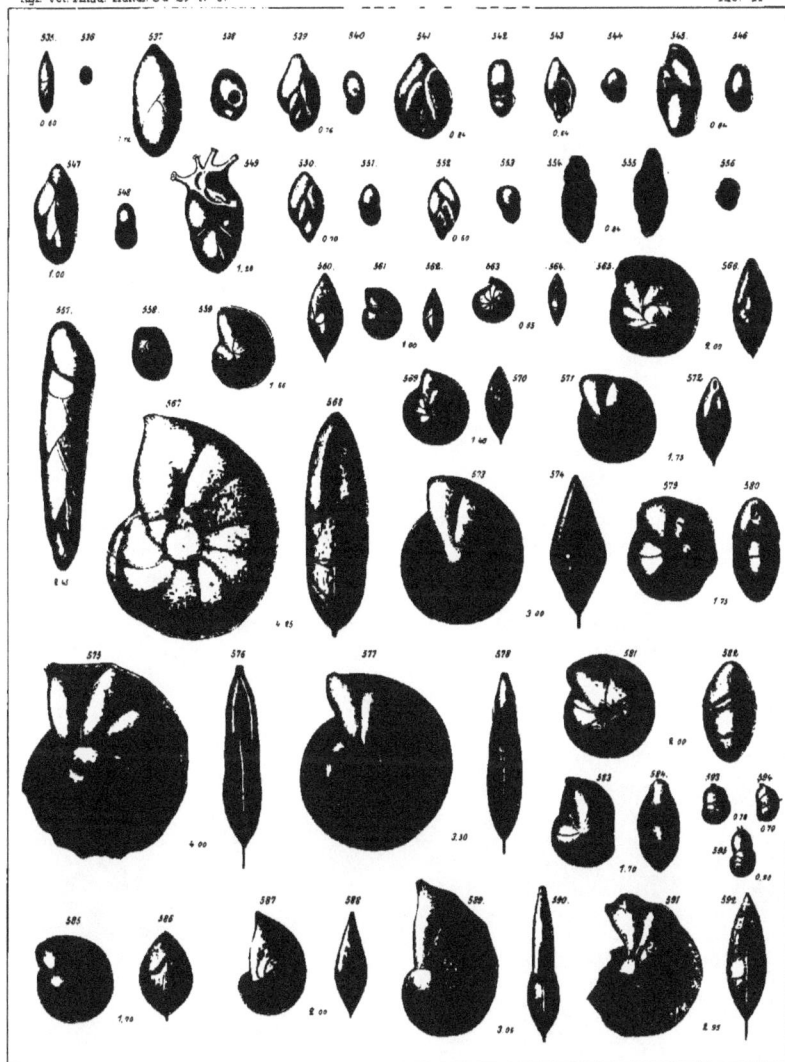

TAB. XI.

Tab. XI.

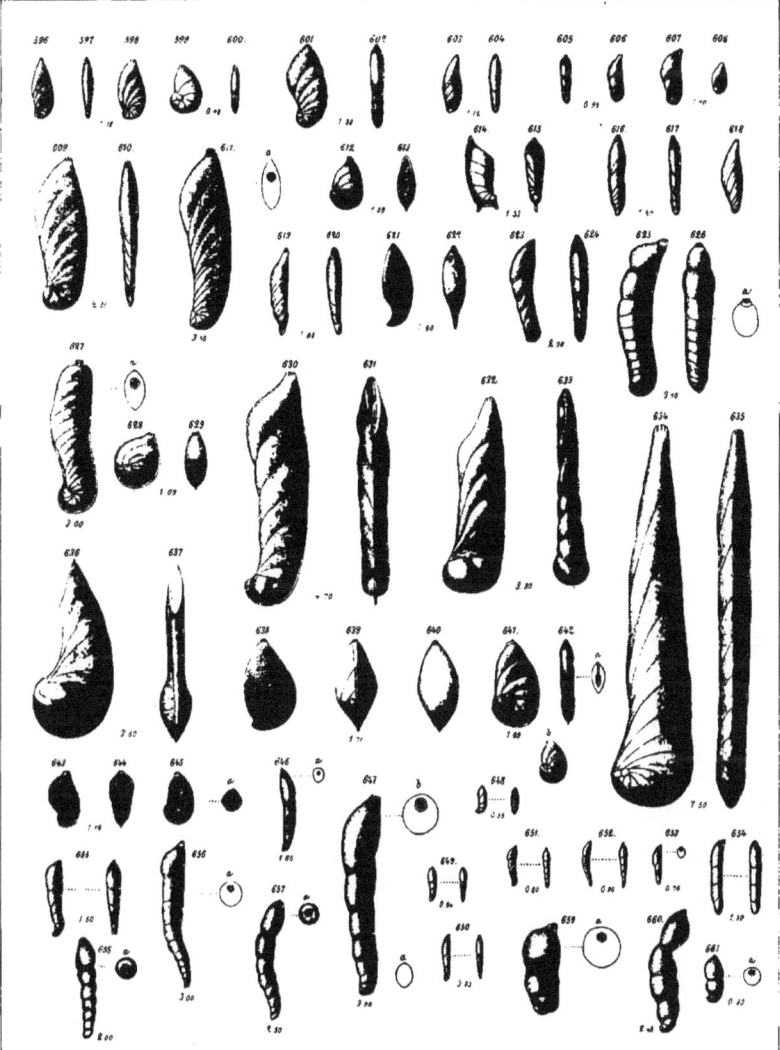

TAB. XII.

Tab. XII.

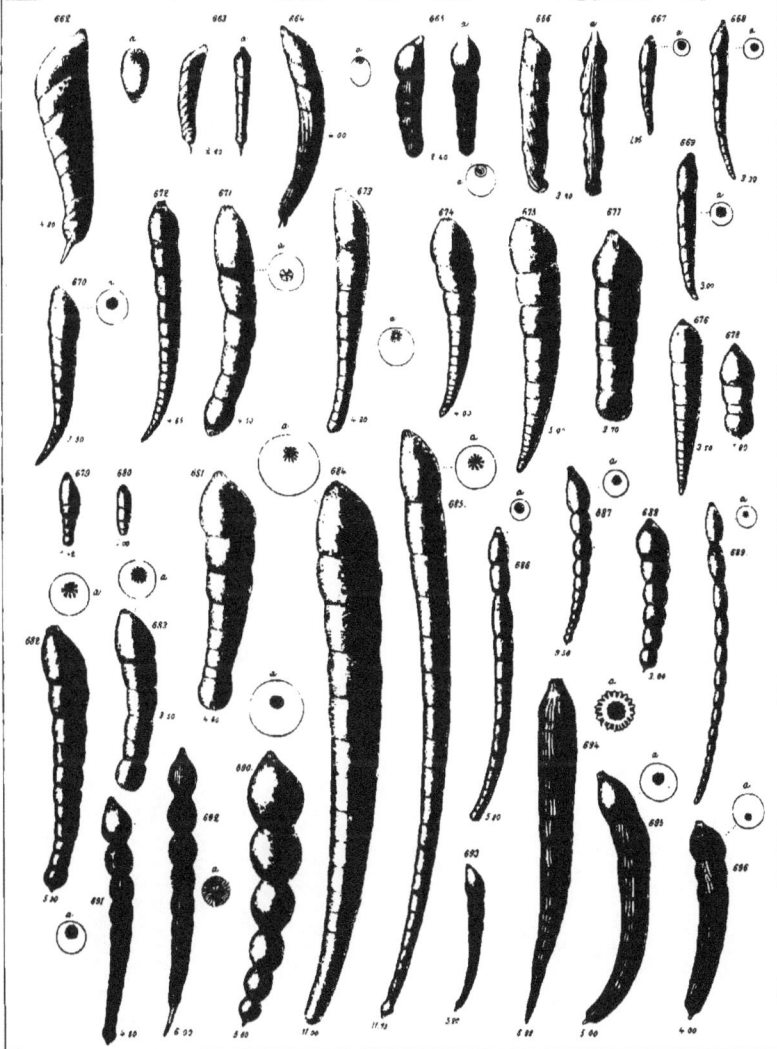

Tab XII

TAB. XIII.

Tab. XIII.

Tab. XIII.

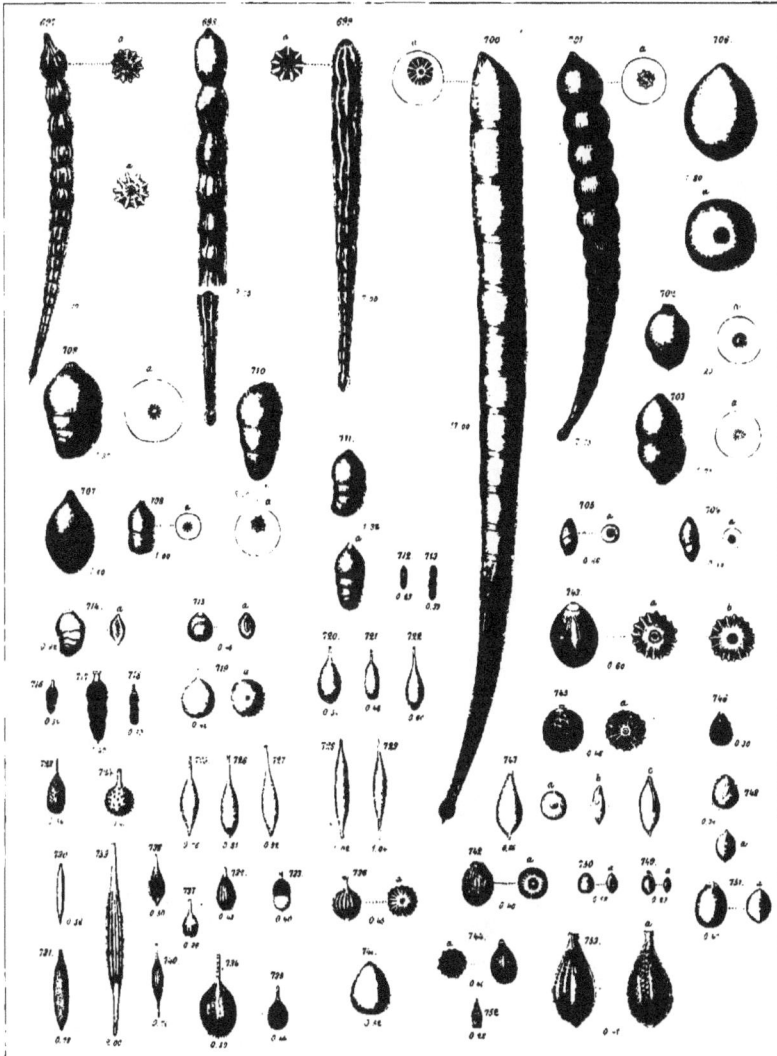

TAB. XIV.

Tab. XIV.

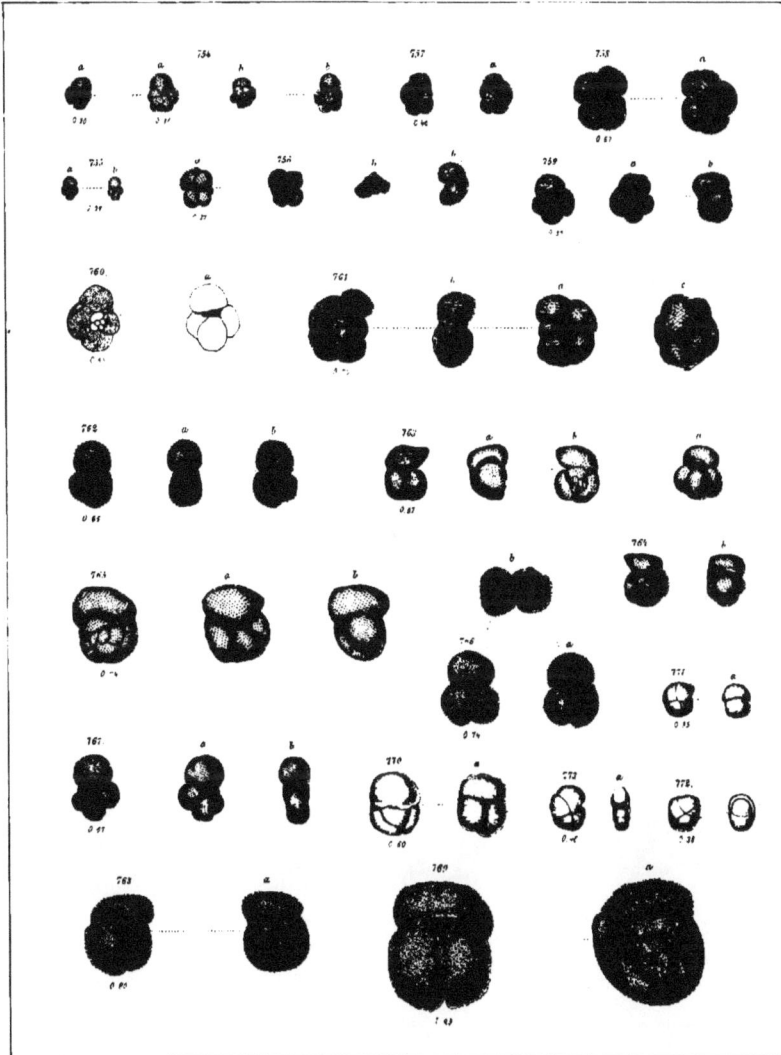

TAB. XV.

Tab. XV.

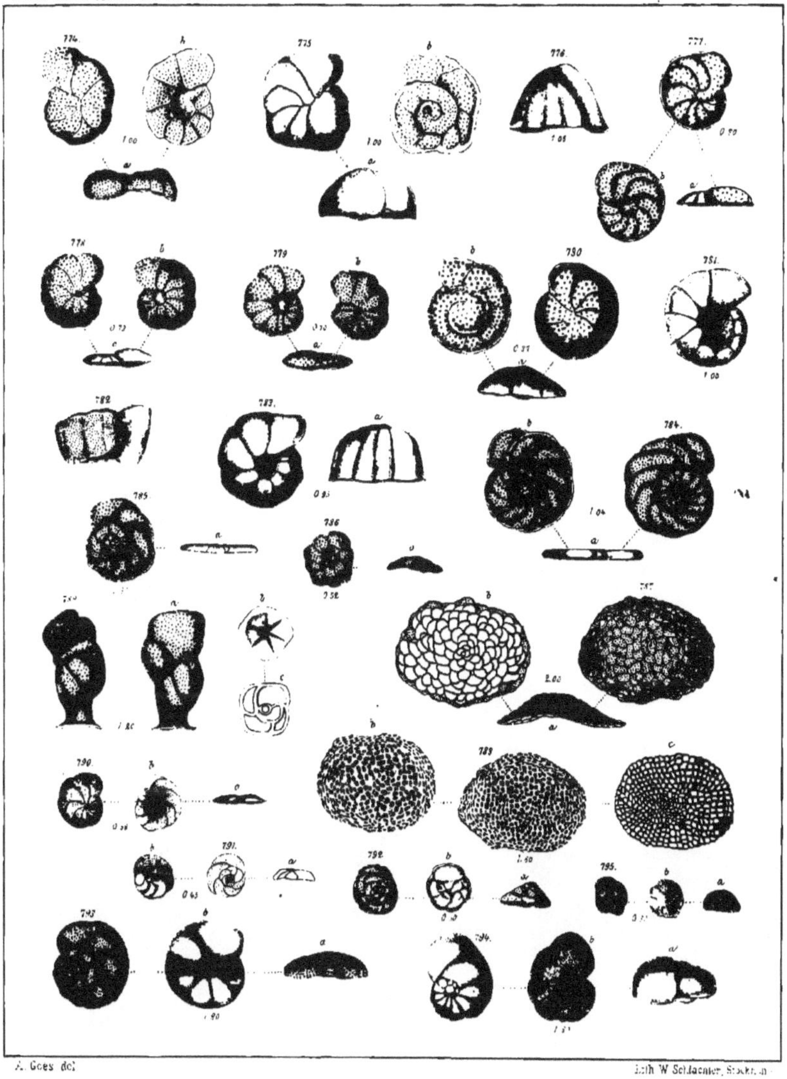

A. Goes del. Lith. W. Schlachter, Stockholm

TAB. XVI.

—

Tab. XVI.

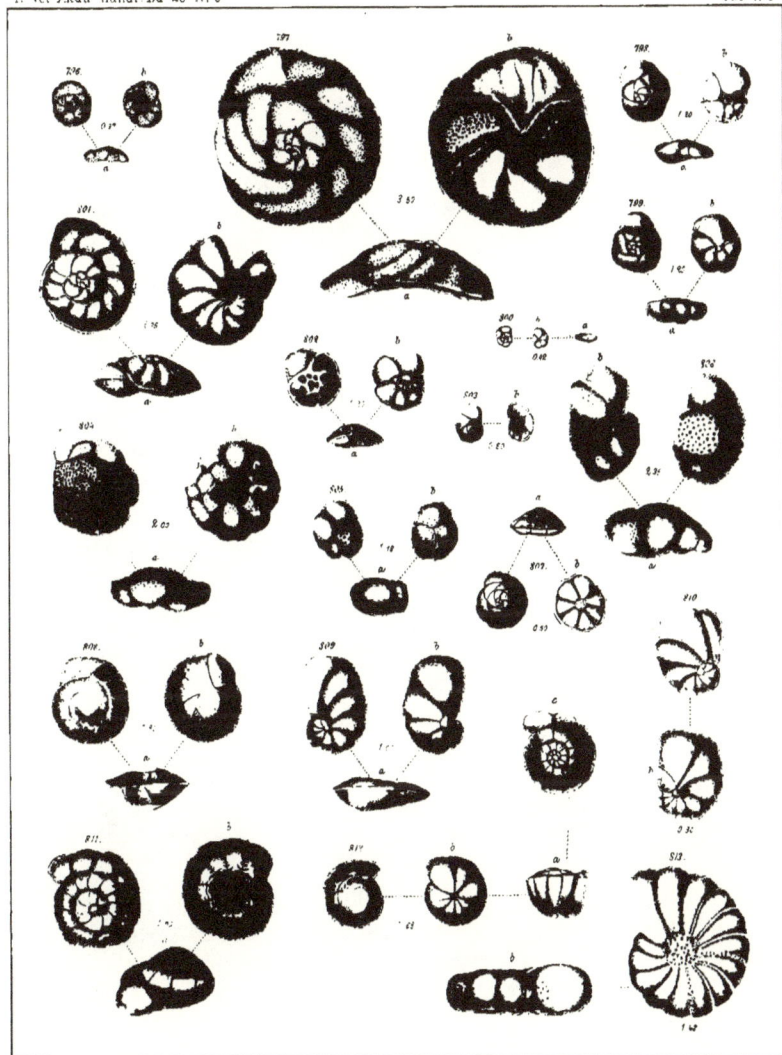

TAB. XVII.

Tab. XVII.

Tab XVII

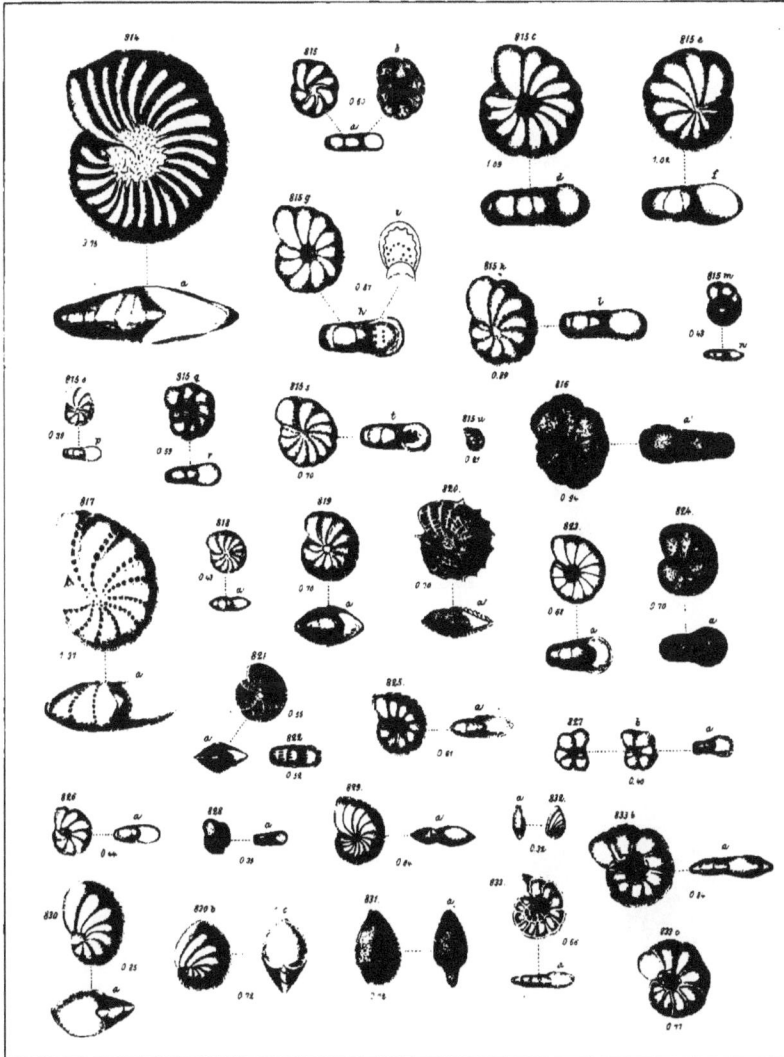

TAB. XVIII.

Tab. XVIII.

Tab XVIII.

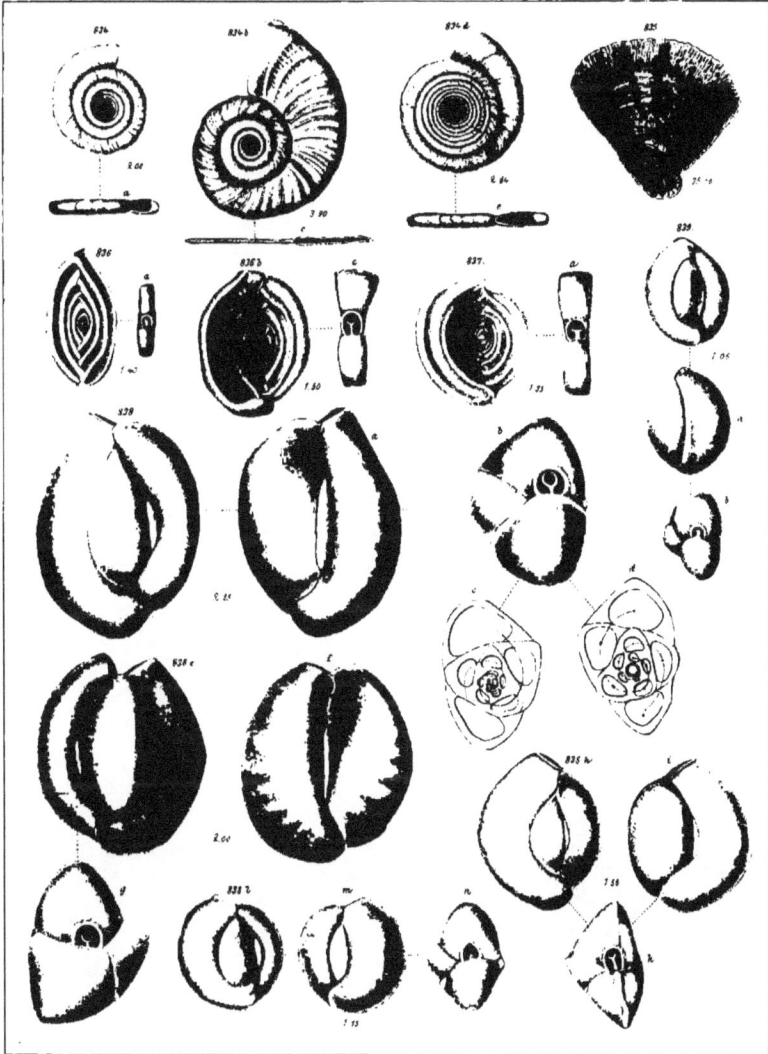

TAB. XIX.

Tab. XIX.

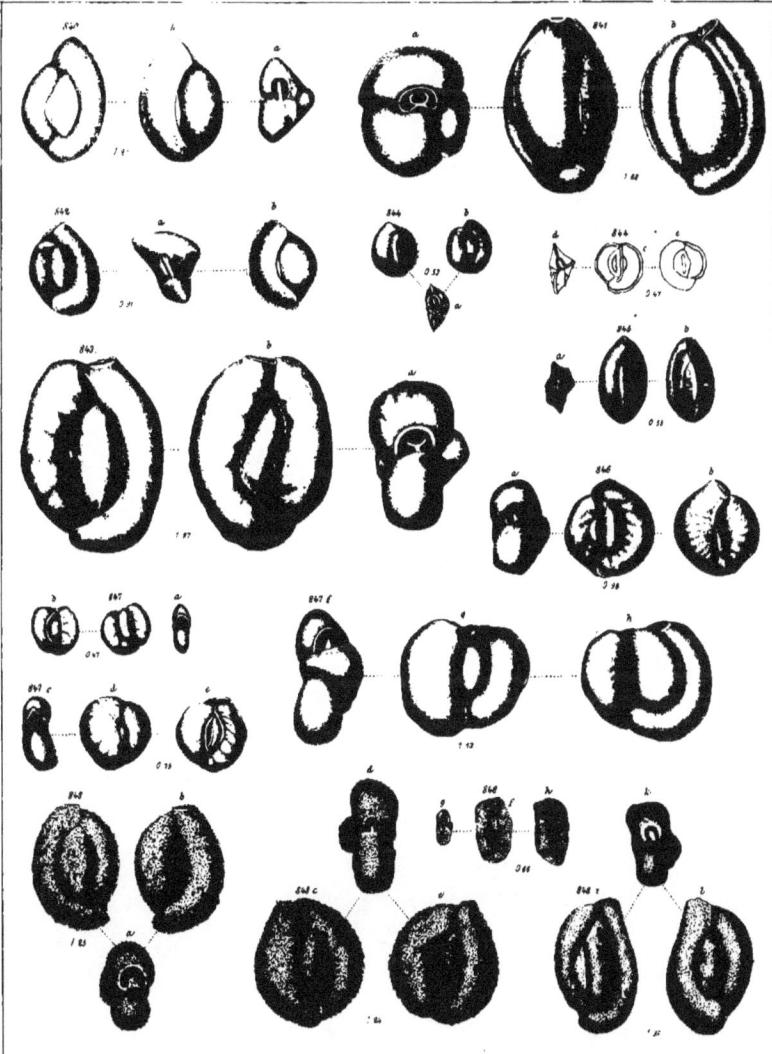

TAB. XX.

Tab. XX.

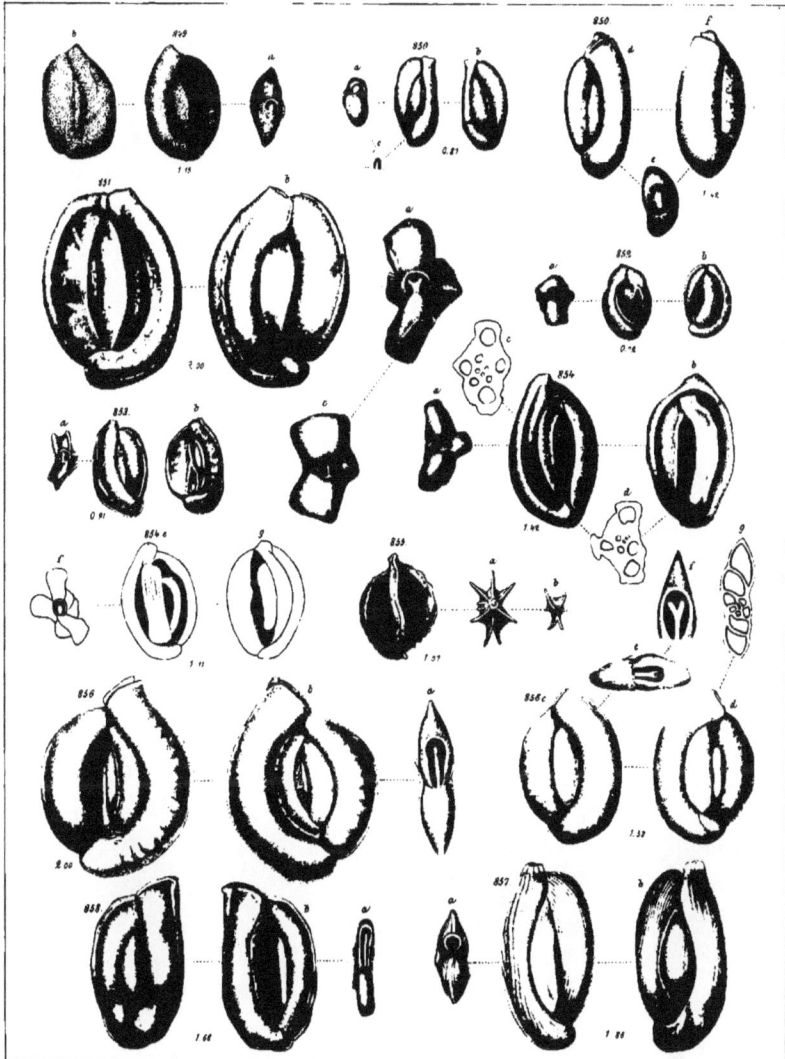

TAB. XXI.

Tab. XXI.

Tab. XXI

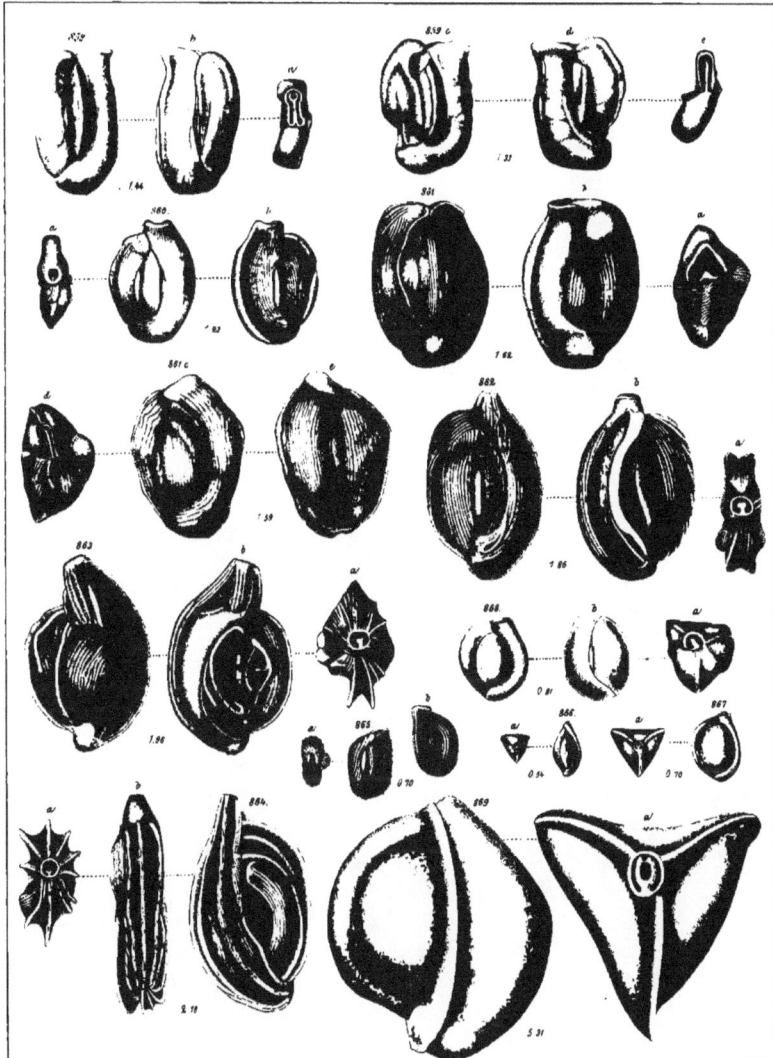

TAB. XXII.

Tab. XXII.

K. Vet. Akad. Handl. Bd. 25. No 9.

Tab. XXII.

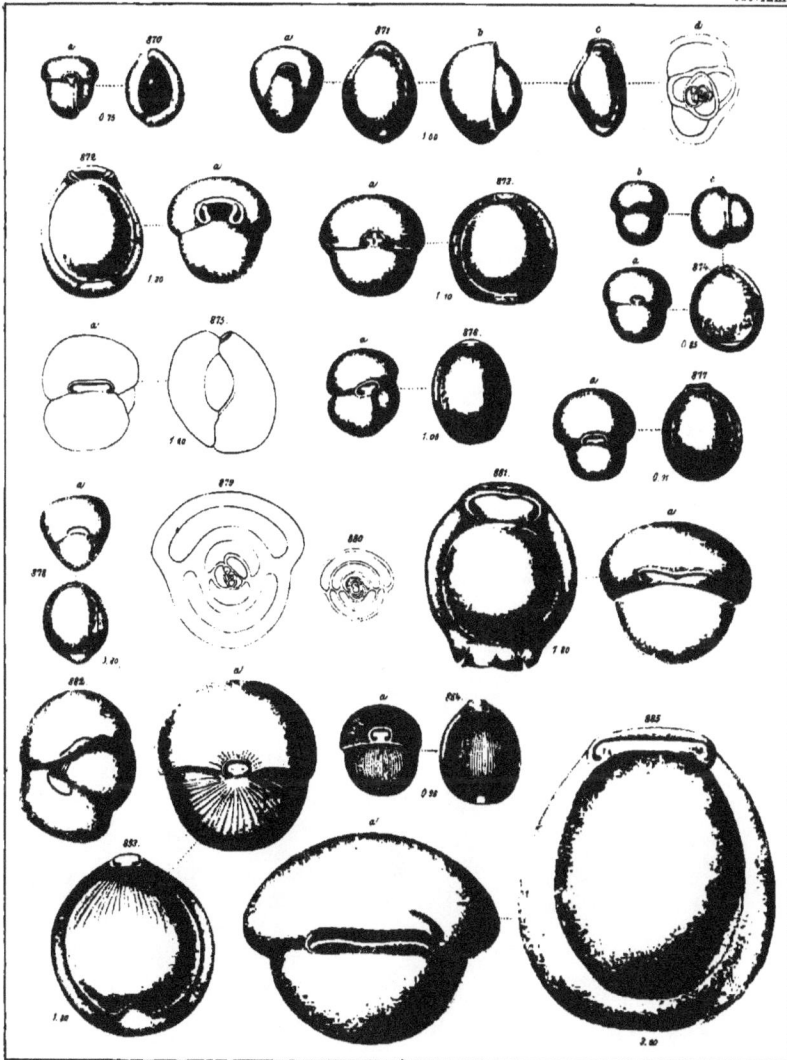

TAB. XXIII.

Tab. XXIII.

Tab XXIII

TAB. XXIV.

Tab. XXIV.

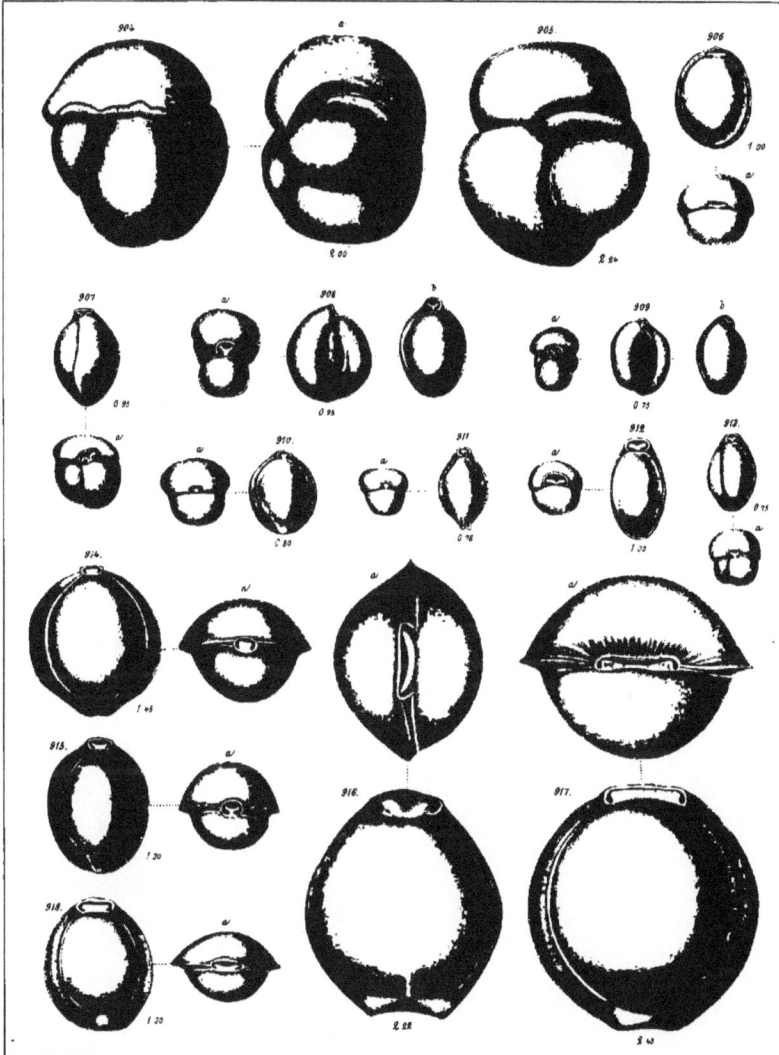

TAB. XXV.

Tab. XXV.

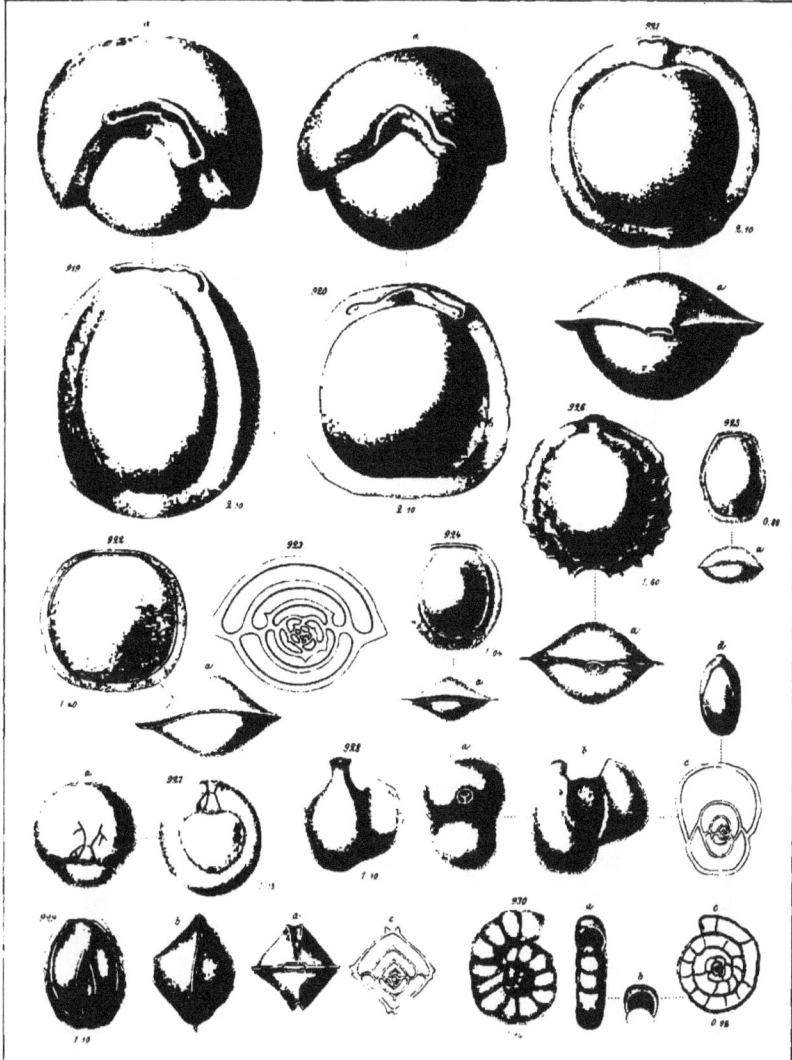